Heine Safety Boiler Co

The Heine safety boiler co.

Manufacturers of water tube steam boilers for all pressures, duties and fuels

Heine Safety Boiler Co

The Heine safety boiler co.

Manufacturers of water tube steam boilers for all pressures, duties and fuels

ISBN/EAN: 9783743467101

Manufactured in Europe, USA, Canada, Australia, Japa

Cover: Foto ©berggeist007 / pixelio.de

Manufactured and distributed by brebook publishing software (www.brebook.com)

Heine Safety Boiler Co

The Heine safety boiler co.

E. D. MEIER,
PRES'T AND CHIEF ENGINEER.

THEO G. MEIER,
V.-P. AND TREAS.

S. D. MERTON,
SECRETARY.

THE HEINE SAFETY BOILER CO.

MANUFACTURERS OF

WATER TUBE STEAM BOILERS

FOR ALL

PRESSURES, DUTIES AND FUELS.

MAIN OFFICE:

ROOMS 703 TO 708 BANK OF COMMERCE BUILDING, No. 421 OLIVE STREET,

ST. LOUIS. MO.

BRANCH OFFICES:

NEW YORK, N. Y.,
120 LIBERTY ST.

PHILADELPHIA, PA.,
669 THE BOURSE.

CHICAGO, ILL.,
1521 MONADNOCK BLDG.

BOSTON, MASS.,
104 EQUITABLE BLDG.

PITTSBURG, PA.,
1212 CARNEGIE BLDG.

REPRESENTATIVES:

DENVER, COLO.,
STEARNS-ROGER MFG. CO.

MONTREAL, CAN.,
GEO. BRUSH,
34 KING ST.

SAN FRANCISCO, CAL.,
RISDON IRON AND LOCO WORKS,
HOWARD AND BEAL STS.

TORONTO, ONT.,
CANADIAN
HEINE SAFETY BOILER CO.

LOUISVILLE, KY.,
R. M. CUNNINGHAM,
612 COLUMBIA BLDG.

ST. LOUIS, MO., AUGUST 1, 1897.

Preface to Second Edition.

IN PRESENTING to the engineering and steam-using world this second and larger edition of "HELIOS," following so closely after the first publication, we wish to express our warm appreciation of the many kind expressions which the first volume has elicited.

The cordial reception given the first edition is in the nature of most distinct and encouraging confirmation of our belief that, in the long run, the best boiler that money can make will find the greatest favor with the greatest number of discriminating steam users.

We submit this second edition to the careful consideration of all who are concerned with the subject of modern boiler practice.

Preface to Fourth Edition.

WE TAKE great pleasure in announcing our fourth edition. We have added an article on "Bagasse" as a boiler fuel, and have entirely rewritten and enlarged our article on "Chimneys and Draft."

We call attention also to seven new and valuable tables, published for the first time in this edition.

HEINE SAFETY BOILER CO.

January 1, 1895.

Preface to Fifth Edition

IN THIS EDITION we desire to call special attention to the revised table of American Coals. The proximate analyses have been omitted, retaining only the heat values and theoretical evaporative powers. The number of tests of coals has been considerably increased.

The article on Fuel Oil has also been considerably enlarged.

July 4, 1896.

"I think 'HELIOS' is immense."

> J. J. DEKINDER,
> Con. Eng. Pennsylvania R. R. Co.

"An invaluable addition to the literature of our profession."

> JOHN L. D. BORTHWICK,
> Chief Engineer U. S. N.

"'HELIOS' throws a brilliant light on many dark subjects."

> JOHN E. CODMAN, C. E. & M. E.
> Philadelphia Water Works.

"It easily takes the lead, even in this age of magnificent catalogues."

> EDWARD K. HILL,
> Prest. Wheelock Engine Co.

"It is a most excellent hand-book, and contains much valuable information."

> R. FORSYTH,
> Eng. Illinois Steel Co.

"'HELIOS,' the most complete book of its kind I have ever seen."

> JOS. H. SPRINGER,
> Gen'l Supt. Frazer & Chalmers.

"I consider 'HELIOS' an excellent addition to my technical library."

> D. ASHWORTH,
> Consulting Engineer.

"It is altogether, to my mind, one of the best publications of its kind that has come out."

> A. J. CALDWELL,
> Hydraulic Engineer.

"It is very well arranged, has a good index, is remarkably free from errors, and is, in fact, just such a book as every engineer should have at hand."

> F. H. BAILEY,
> Chief Engineer U. S. N.

"The data being the result of recent experiment and experience, furnishes a fund of information not found in other text books, and is a valuable addition to mechanical literature."

> F. S. ALLEN,
> Chief Inspector, Hartford Steam Boiler Insp. & Ins. Co.

"Your beautiful contribution to the technical literature of the day was on my desk on my return from Chicago. It is one of the handsomest bits of its kind that has yet appeared, and you are to be heartily congratulated on your success."

> R. H. THURSTON,
> Prof. Mech. Engr. Cornell University.

HELIOS.

Source of All Power! Fountain of Light and Warmth!!

Adored by the ancient husbandman as the God who blessed his labors with a harvest of golden grain; revered by the early sage as the great visible means of the divine creative force; pictured by the inspired artist as the tireless charioteer who drives his four fiery steeds daily across the heavens, his head circled by a crown of rays his chariot wheel the disk of the sun itself.

When primeval man began to think, the sun seemed to him the cause of all those wonders in nature which ministered to his simple wants, or taught his soul to hope. His crude feelings of awe and gratitude blossomed into worship, and we find the sun as central figure in all early religions. He was the Suraya of the Hindoos, the Baal of the Phœnicians, the Odin of the Norsemen, and his temples arose alike in ancient Mexico and Peru. As Mithras of the Parsees, he was adored as the symbol of the Supreme Deity, his messenger and agent for all good. As Osiris he received the worship and offerings of the Egyptians, whose priests, early adepts in the rudiments of science, saw in him the cause of the annual fructifying overflow of the Nile.

Modern knowledge, with its vast array of facts and figures, can but verify and seal the faith of these ancient observers. What they dimly discerned as probable is now the central fact of physical science. From him are derived all the forces of nature which have been yoked into the service of man. All animal and plant life draws its daily sustenance from the warmth and light of the sun, and it is but his transmuted energy we expend, when, with muscle of man or horse, we load our truck or roll it along the highway. Do we irrigate the soil from the pumps of a myriad windmills? His rays, on plains far inland, supply the energy for the breeze which turns their vanes!

Does a lumbering wheel drive a dozen stamps and a primitive arastra in some Mexican canyon? Do mighty turbines whirl a million flying spindles and shake thousands of clattering looms on the banks of some New England stream? From the bosom of the ocean and the swamps of the tropics, Helios lifted those vapory Titans whose lifeblood courses in the mountain torrent and the river of the plain!

Do a hundred cars rattle up the steep streets of the smiling city by the Golden Gate? Are massive ingots of steel forged to shape and size by the giant hammers of Bethlehem? The fuel which gives them motion was stored for us, ages before man was evolved, by the rays which flash from his chariot wheels! "The heat now radiating from our fire places has at some time previously been transmitted to the earth from the sun. If it be wood that we are burning, then we are using the sunbeams that have shone on the earth within a few decades. If it be coal, then we are transforming to heat the solar energy which arrived at the earth millions of years ago."

Professor Langley remarks that "the great coal fields of Pennsylvania contain enough of the precious mineral to supply the wants of the United States for a thousand years. If all that tremendous accumulation of fuel were to be extracted and burned in one vast conflagration, the total quantity of heat that would be produced would, no doubt, be stupendous, and yet," says this authority, who has taught us so much about the sun, "all the heat developed by that terrific coal fire would not be equal to that which the sun pours forth in the thousandth part of each single second."

The almost limitless stores of petroleum which are found in America and in Asia, and the smaller, though still vast supplies of natural gas which some favored localities are now exploiting, represent but so much sun-energy transmuted through forests of prehistoric vegetation.

Another authority tells us that the total amount of living force "which the sun pours out yearly upon every acre of the earth's surface, chiefly in the form of heat is 800,000 horse-power." And he estimates that a flourishing crop utilizes only $\frac{4}{10}$ of 1 per cent of this power.

Remembering, then, that this sun-energy reaches us only one-half of each day, we may, *whenever we learn how*, pick up on every acre an average of 175 horse-power during each hour of daylight, as a surplus which nature does not require for her work of food production.

Attempts to utilize this daily waste have been made, and future inventors may fire their boilers directly with the radiant heat of the sun. But whether we depend on what he garnered for us ages ago, or quite recently, or on the stores he will lavish on us in the future, it is clear that man's continued existence on earth is directly dependent on Helios.

In olden times the various trades or guilds chose as their patron saint some prominent person who was thought to have embodied in his life-work the special means and methods of their craft. By that token we claim Helios as our own. He has always carried the record for evaporative efficiency. He provides both the fuel and the water for our boilers. He teaches us perfect circulation, upward as mingled vapor and water by the action of heat, and down again by gravity as rain and river in solid water. It is therefore fit that the boiler in which this perfect and unobstructed circulation is made the leading feature of construction should have HELIOS as its emblem!

In the following pages we give some account of the fuels used in the practical arts, of the water which becomes the vehicle for transmitting their energy into mechanical power, and of the limitations imposed by their varying conditions. These must all be taken into account in estimating how much we may expect of certain combinations of machinery. Much of the text and many of the tables are taken from Mr. David Kinnear Clark's admirable book on the steam engine, for which his consent and that of his publishers, Messrs. Blackie & Son, was courteously given. We also, by permission, quote freely from such authorities as Mr. Emerson McMillin, Prof. Wm. B. Potter, Prof. R. H. Thurston, Mr. J. M. Whitham, Prof. D. S. Jacobus, Prof. Ordway and others. Thanks are also due for valuable matter to Messrs. Henry R. Worthington, The B. F. Sturtevant Co., Mr. Alfred R. Wolff, Mr. C. W. Owston and Messrs. Hunt & Clapp. In most instances we indicate the scource by initials.

We trust that the tables and data may be found convenient for ready reference alike by professional men, by manufacturers, and by that growing class of practical steam engineers who realize that true theory, consonant with collective experience, is within the reach of every thoughtful man who pulls the throttle.

E. D. M.

Main Boiler Plant of World's Columbian Exposition, CHICAGO, ILL. Showing 3000 H. P. Plant of Heine Boilers.

HEAT.

Heat is the form in which we receive most of the sun-energy. In the various fuels it exists in a potential form requiring combustion, *i. e.*, combination of the active elements of the fuel with the oxygen of the air, to reappear in its active form.

"HEAT AS A FORM OF ENERGY is subject to the general laws which govern every form of energy and control all matter in motion, whether that motion be molecular or the movement of masses.

"That heat is the motion of the molecules of bodies was first shown by experiment by Benjamin Thompson, Count Rumford, then in the service of the Bavarian Government, who in 1798 presented a paper to the Royal Society of Great Britain, describing his work, and reciting the results and his conclusion that heat is not substance, but a form of energy.

"This paper is of very great historical interest, as the now accepted doctrine of the persistence of energy is a generalization which arose out of a series of investigations, the most important of which are those which resulted in the determination of the existence of a definite quantivalent relation between these two forms of energy and a measurement of its value, now known as the 'mechanical equivalent of heat.' The experiment consisted in the determination of the quantity of heat produced by the boring of a cannon at the arsenal at Munich."

Work in the same direction was done by Sir Humphrey Davy, Sadi Carnot, Dr. Mayer and Mr. Colding. But Dr. Joule, from 1843 to 1849, made a series of experiments by various methods, the results of which have been generally accepted as satisfactory.

Quantities of heat are measured, in English units, by what is termed the British Thermal Unit, or for brevity, B. T. U. The B. T. U. is the quantity of heat required to raise 1 lb. of pure water from a temperature of 62° F. to 63° F., and has an equivalent in mechanical units of work. This is frequently called simply a Heat Unit or designated by H. U.

The mechanical unit of work is the *foot-pound*, or the work required to raise 1 pound, 1 foot high. Joule's experiments, and those of later investigators, show 778 ft. lbs. to be equivalent to one B. T. U. This number, 778, is known as Joule's equivalent or symbolically J. 33000 ft. lbs. per min. was called a *horse power* by Watt, and is used as such to-day, it being the unit for large powers.

The electrical unit of power is the *Watt*, which is the product of 1 ampere × 1 volt. 746 Watts are equivalent to 1 H. P. or 33000 ft. lbs. Hence the Watt has an equivalent in heat units also.

Water power is measured in terms of the height of fall or velocity of flow, and the quantity or weight of water passing, the result, however, being in mechanical units. Hence $P = H \times W \times V$, where P = ft. lbs. per sec., H = height of fall in ft., W = weight per cu. ft. of water, V = cubic feet of water falling per second.

Since $v^2 = 2\,gH$. we have $P = \frac{v^2}{2g} \times V \times W$ where P, V, and W, are the same as before and v the velocity of flow of the water in ft. per sec. and $g = 32.2$.

Owing to the frictional losses and the inefficiency of all kinds of water motors, more than 80 per cent. of this theoretical power is rarely ever realized. The best types of water motors give only 80 to 90 per cent. efficiency.

The following table shows the relation of the various units :

TABLE NO. 1.

Equivalents of Work and Heat.

B. T. U.		Ft. lbs.		Watts.		
1	=	778	=	17.59		
42.41	=	33000	=	746	=	1 H. P.

In the French or metric system of units, a Heat Unit or *Calorie* is the quantity of heat required to raise 1 Kilogram of pure water $1°$ Cent. at or about $4°$ C.

The following tabular statement shows the relation of the French and English units :

TABLE NO. 2.

French and English Units Compared.

1 Calorie...3.968 B. T. U.
0.252 Calorie...1 B. T. U.
French Mechanical Equivalent, }
425.0 Kilogram-metres, } ....3075 ft. lbs.
107.7 Kilogram-metres............... ....J, or 778 ft. lbs.

For convenience in translating French or German results in to English or American we have the following compound units :

TABLE NO. 3.

Equivalent Compound Units.

1 Calorie per square metre_____ _____0.369 B. T. U. p. square ft.
1 B. T. U. or 1 H. U. p. square ft_____2.713 Cal. p. square metre.
1 Calorie p. Kilogram_____1.800 H. U. per pound.
1 H. U. p. pound_____0.556 Cal. p. Kilogram.

"HEAT TRANSFORMATIONS may take place, through the action of physical and chemical forces, into any other known form of energy, and another form of energy may be transmuted into heat. Nearly all physical phenomena, in fact, involve heat-transformation in one form or another, and in a greater or less degree, under the laws of energetics. According to the first of those laws, such changes must always occur by a definite quantivalence, and when heat disappears in known quantity it is always certain that energy of calculable amount will appear as its equivalent ; the reverse is as invariably the case when heat is produced ; it always represents and measures an equivalent amount of mechanical, electrical, chemical, or other energy.

"Heat and Mechanical Energy are thus evidently subject to the general laws of transformation of energy, and the transmutation of the one into the other must always be capable of treatment mathematically. The relations of these two forms of energy are taken as the subject of a division of energetics known as the science of thermodynamics, and a vast amount of study and research has been given by the ablest mathematical physicists of modern times to the investigation of its laws and their applications, and to the building up of that science.

"The conversion of water into steam in the steam boiler and the utilization of the heat-energy thus made available, or in heated air and other gases, in steam or other heat-engines, constitute at once the most familiar and the most important of known illustrations of thermodynamic phenomena and their useful application. The process of making steam is one of production of heat by transformation from the potential form of energy through the action of chemical forces, and its storage in sensible form for later use in the steam-engine, where it is changed into equivalent mechanical energy. The pure science of the steam-engine is thus the science of thermodynamics, the first applications of which are made in the operations carried on in the steam-boiler.

"SENSIBLE AND LATENT HEATS must be carefully distinguished in studying the action of heat on matter. The term 'Sensible Heat' scarcely requires definition ; but it may be said that sensible and latent heats represent latent and sensible work ; that the former is actual, kinetic, heat-energy, capable of transformation into mechanical energy, or vis viva of masses, and into mechanical work ; while the latter form is not heat, but is the equivalent of heat transformed to produce a visible effect in the performance of molecular, or internal as well as external, work, and visible alteration of volume and other physical conditions.

"It is seen that heat may become 'latent' through any transformation which results in a definite and defined physical change, produced by expansion of any substance in consequence of such transmutation into internal and external work ; whether it be simple increase of volume or such increase with change of physical state.

"THE LATENT HEAT OF EXPANSION is a name for that heat which is demanded to produce an increase of volume, as distinguished from that untransformed heat which is absorbed by the substance to produce elevation of temperature. The latent heat of expansion may, by its absorption and transformation, and the resulting performance of internal and external work, cause no other effect than change of volume, as e. g., when air is heated, or it may at the same time produce an alteration of the solid to the fluid, or of the liquid to the vaporous state, as in the melting of ice or the boiling of water, in which latter cases, as it happens, no elevation of temperature occurs, all heat received being at once transformed. In the expansion of air, and in other cases in which no such change of state occurs, a part of the heat absorbed remains unchanged, producing elevation of temperature ; while another part is transformed into latent heat of expansion."

R. H. T.

We give below tables of the boiling and melting points of various substances, and the linear expansion of various solids.

Boiling Points of Various Substances.

At Atmospheric Pressure at Sea Level.

SUBSTANCE.	Degrees Fahr.	SUBSTANCE.	Degrees Fahr.
Alcohol	173	Sulphur	570
Ammonia	140	Sulphuric Acid, s. g. 1.848	590
Benzine	176	Sulphuric Acid, s. g. 1.3	240
Coal Tar	325	Sulphuric Ether	100
Linseed Oil	597	Turpentine	315
Mercury	648	Water	212
Naptha	186	Water, Sea	213.2
Nitric Acid, s. g. 1.42	248	Water, Saturated Brine	226
Nitric Acid, s. g. 1.5	210	Wood Spirit	150
Petroleum Rectified	316		

Melting Points of Metals.

From D. K. C.

Melting Points of Various Solids.

From D. K. C. and H.

METAL.	Degrees Fahr.	SUBSTANCE.	Degrees Fahr.
Aluminum	Full Red Heat.	Carbonic Acid	—108
Antimony	1150	Glass	2377
Bismuth	507	Ice	32
Bronze	1690	Lard	95
Copper	1996	Nitro-Glycerine	45
Gold, Standard	2156	Phosphorus	112
Gold, Pure	2282	Pitch	91
Iron, Cast, Gray	2012	Saltpetre	606
	1922	Spermaceti	120
Iron, Cast, White	to		109
	2012	Stearine	to
Iron, Wrought	2912		120
Lead	617	Sulphu.	239
Mercury	—39	Tallow	92
Silver	1873	Turpentine	14
	2372	Wax, Rough	142
Steel	to	Wax, Bleached	154
	2552		
Tin	442		
Zinc	773		

Melting Points of Fusible Plugs.

From D. K. C.

	Softens at	Melts at		Softens at	Melts at
2 Tin, 2 Lead	365	372	2 Tin, 7 Lead	377½	388
2 Tin, 6 Lead	372	383	2 Tin, 8 Lead	395½	408

Expansion of Solids at Ordinary Temperatures.

D. K. C.

SUBSTANCE.	Coefficient for 1° Fahr.	Total Expansion between 32° Fahr. and 212° Fahr.		In Length of 10 Feet.	
		Coefficient.			
		Decimal.	Fraction.		
				Feet.	Inches.
Aluminum (Cast)	.00001234	.002221	1/450	.02221	.2664
Antimony (Crystallized)	.00000627	.001129	1/885	.01129	.1336
Brass (Cast)	.00000957	.001723	1/581	.01723	.2067
Brass (English Plate)	.00001052	.001894	1/529	.01894	.2273
Brass (Sheet)	.00001040	.001872	1/535	.01872	.2246
Brick (Best Stock)	.00000306	.000550	1/1818	.00550	.0660
Brick in Cement Mortar (Headers)	.00000494	.000890	1/1123	.00890	.1068
Brick in Cement Mortar (Stretchers)	.00000256	.000460	1/2174	.00460	.0552
Bronze	.00000975	.001755	1/568	.01755	.2106
Cement (Roman, Dry)	.00000797	.001435	1/694	.01435	.1722
Cement (Portland, Neat)	.00000594	.001070	1/935	.01070	.1284
Cement (Portland, with Sand)	.00000656	.001180	1/847	.01180	.1416
Copper	.00000887	.001596	1/625	.01596	.1915
Glass (Flint)	.00000451	.000812	1/1234	.00812	.0974
Glass (White, Free from Lead)	.00000492	.000886	1/1130	.00886	.1063
Glass (Blown)	.00000498	.000896	1/1111	.00896	.1075
Glass (Thermometer)	.00000499	.000897	1/1111	.00897	.1076
Glass (Hard)	.00000397	.000714	1/1400	.00714	.0857
Granite (Gray, Dry)	.00000438	.000789	1/1266	.00789	.0947
Granite (Red, Dry)	.00000498	.000897	1/1111	.00897	.1076
Gold (Pure)	.00000786	.001415	1/707	.01415	.1698
Iron (Wrought)	.00000648	.001166	1/866	.01166	.1399
Iron (Swedish)	.00000636	.001145	1/873	.01145	.1374
Iron (Cast)	.00000556	.001001	1/1000	.01001	.1201
Iron (Soft)	.00000626	.001126	1/897	.01126	.1351
Lead	.00001571	.002828	1/355	.02828	.3394
Marble (Ordinary, Dry)	.00000363	.000654	1/1530	.00654	.0785
Marble (Ordinary, Moist)	.00000663	.001193	1/838	.01193	.1432
Mercury (Cubic Expansion)	.00009984	.017971	1/56	.17971	2.1565
Nickel	.00000695	.001251	1/800	.01251	.1501
Plaster (White)	.00000922	.001660	1/602	.01660	.1992
Platinum	.00000479	.000863	1/1159	.00863	.1036
Silver (Pure)	.00001079	.001943	1/514	.01943	.2334
Slate	.00000577	.001038	1/967	.01038	.1246
Steel (Cast)	.00000636	.001144	1/874	.01144	.1373
Steel (Tempered)	.00000689	.001240	1/806	.01240	.1488
Stone (Sand, Dry)	.00000652	.001174	1/852	.01174	.1409
Tin	.00001163	.002094	1/477	.02094	.2513
Wood (Pine)	.00000276	.000496	1/2016	.00496	.0595
Zinc	.00001407	.002532	1/395	.02532	.3038
Zinc 8, Tin 1	.00001496	.002692	1/372	.02692	.3230

The Specific Heat of a body signifies its capacity for heat or the quantity of heat required to raise the temperature of the body one degree Fahrenheit, compared with that required to raise the temperature of an equal weight of water one degree.

TABLE NO. 7.

Specific Heats.

D. K. C.

SUBSTANCE.	SPECIFIC HEAT.	SUBSTANCE.	SPECIFIC HEAT.
Ice	0.504	Anthracite	0.2017
Water at 32° F	1.000	Oak Wood	0.570
Gaseous Steam	0.475	Fir Wood	0.650
Saturated Steam	0.305	Oxygen (Equal	
Mercury	0.0333	Weights; Constant Volume)	0.1559
Sulphuric Ether, Density .715	0.5200	Air (at Constant Pressure)	0.2377
Alcohol	0.6588		
Lead	0.0314	Air (Equal Weights Constant Vol.)	0.1688
Gold	0.0324		
Tin	0.0566	Nitrogen (Equal Wgts; Constant Volume)	0.1740
Silver	0.0570		
Brass	0.0939		
Copper	0.0951	Hydrogen (Equal Wgts; Constant Volume)	2.4096
Zinc	0.0956		
Nickel	0.1086		
Wrought Iron	0.1138 to 0.1255	Carbonic Oxide (Equal Weights; Constant Vol.)	0.1768
Steel	0.1165 to 0.1185		
Cast Iron	0.1298		
Brickwork and Masonry	0.200	Carbonic Acid (Equal Weights; Constant Vol.)	0.1714
Coal	0.2411		

Boiler Plant of Lannett Cotton Mills,
WEST POINT, GA.
900 H. P. Heine Boilers.

The Grand Republic Mills of the Russell & Miller Milling Co.,
WEST SUPERIOR, WIS.

Contains at present 500 H. P. Plant of Heine Boilers, with room for future extension to 2000 H. P.

COMBUSTION.

Combustion or Burning is the chemical combination of the constituents of the fuel, mostly carbon and hydrogen, with the oxygen of the air. The nitrogen remains inert and causes loss of useful effect to the extent of the heat it carries off through the chimney.

The hydrogen combines with enough oxygen to form water which passes off as steam.

The carbon combines with enough oxygen to form carbonic acid or carbon dioxide gas (perfect combustion) or with only enough to form carbonic oxide or carbon monoxide gas (imperfect combustion).

The following table gives the quantities of air, the heat evolved and the resulting temperature from the combustion of constituent parts of fuel, under the supposition that the chemical requirements are exactly fulfilled:

TABLE NO. 8.

Combustion Data.

O. H. I..

COMBUSTIBLE.	Atomic Weight.	COMBUSTION PRODUCT.	Wgt. of Oxygen per lb of Combustible	Amount of air consumed per lb. of combustible.		Calorific power. Heat units p.lb.of combustible	Resulting temperat'e of combustion. No surplus air assumed.
	(H₂ I)		Lbs.	Lbs.	cu. ft. 62° F.	B.T.U.	Deg. Fahr.
Oxygen (O)	16						
Hydrogen (H)	1	Water (H₂O)	8.0	34.8	457	62032	5898
Carbon (C)	12	Carbonic oxide (CO)	1.33	5.8	76	4452	2358
Carbon (C)	12	Carbon dioxide (CO₂)	2.66	11.6	152	14500	4939
Carbonic oxide (CO)	28	Carbon dioxide	0.57	2.48	33	4325	5508
Marsh gas (C H₄) (light hydrocar'n)	16	CO₂ and H₂O	4.00	17.4	229	26383	9624
Olefiant gas (C₂H₄) (heavy hydrocarbon)	24	CO₂ and H₂O	3.43	15.0	196	21290	9775
Sulphur (S)	32	SO₂	1.00	4.35	57	4032	3637

Conditions for the Complete Combustion of Fuel in Furnaces.

For insuring completeness of combustion, the first condition is a sufficient supply of air ; the next is that the air and the fuel, solid and gaseous, should be thoroughly mixed ; and the third is that the elements—air and combustible gases—should be brought together and maintained at a sufficiently high temperature. The hotter the elements the greater is the facility for good combustion.

RULE 1. *To find the quantity of air at 62° F., under one atmosphere, chemically consumed in the complete combustion of one pound of fuel of a given composition.* Let the constituent carbon, hydrogen, and oxygen be expressed as percentages of the total weight of the fuel. To the carbon add three times the hydrogen, and from the sum deduct four-tenths of the oxygen. Multiply the remainder by 1.52. The product is the quantity of air at 62° F. in cubic feet.

Formula :—$A = 1.52 (C + 3 H — .4O)$ (1)

To find the weight of the air chemically consumed, divide the volume found as above by 13.14 ; the quotient is the weight of the air in pounds.

RULE 2. *To find the total weight of the gaseous products of the complete combustion of one pound of a fuel*, multiply the percentage of constitutent carbon in the fuel by 0.126, and that of hydrogen by 0.358. The sum of these products is the total weight of the gases in pounds.

$$\text{Formula :--} W = 0.126 \ C + 0.358 \ H \quad (2)$$

RULE 3. *To find the total volume, at 62° F., of the gaseous products of the complete combustion of one pound of fuel*, multiply the constituent percentage of carbon in the fuel by 1.52, and that of hydrogen by 5.52. The sum of these products is the total volume in cubic feet.

$$\text{Formula :--} V = 1.52 \ C + 5.52 \ H \quad (3)$$

The corresponding volume of the gases at other temperatures is given by the formula—

$$V' = V \tfrac{t' + 461}{523} \quad (4)$$

In which V is the volume at 62° F., t' is the other temperature and V' the corresponding volume. That is to say, the volume at any other temperature t' is found by multiplying the volume at 62° by (t' plus 461), and dividing by 523.

RULE 4. *To find approximately the total heating power of one pound of a combustible, of which the percentages of the constituent carbon and hydrogen are given.* To the carbon add 4.28 times the hydrogen, and multiply the sum by 145. The product is the heating power in British units.

$$\text{Formula :--} h = 145 \ (C + 4.28 \ H) \quad (5)$$

RULE 5. *To find the total evaporative power, at 212° F., of one pound of combustible, of which the percentages of the constituent carbon and hydrogen are given.* To the carbon add 4.28 times the hydrogen, and multiply the sum by 0.13 when the water is supplied at 62° F., or by 0.15 when the water is supplied at 212° F. The product is the total evaporative power of one pound of the combustible, in pounds of water evaporated at 212° F.

$$\text{Formula :--(Water supplied at 212°),} E = 0.15 \ (C + 4.28 \ H) \quad (6)$$

In most cases an additional or surplus quantity of air is required to facilitate the completion of the combustion of fuel beyond that which is chemically consumed ; *but the proportion of surplus air required appears to diminish as the rate of combustion and the general temperature in the furnace are increased.* For instance, for the most perfectly managed furnaces of boilers in Cornwall, Mr. Hunt found that there was as much free oxygen in the gaseous products in the chimney as was chemically consumed in combustion. Again, Messrs. Delabeche and Playfair found that the surplus oxygen varied from a fourth to a half; and from the statements of Mr. Longridge with regard to experimental trials at Newcastle with Hartley coal, it appears that the surplus air amounted to only 9 per cent. These proportional surplus quantities, observed under very different circumstances, are found to diminish as the rates of combustion increase, thus:

Coal Consumed per Square Foot
of Grate per Hour.   Surplus Air.

Cornish Boilers............ 2 lbs. to 4 lbs ......100 per cent.
Delabeche and Playfair...10 lbs. to 16 lbs ...... 25 to 50 per cent.
Longridge..................20 lbs. and upwards.. 9¾ per cent.

These results are roughly indicative of the law of the excess of air.   In the instance of Hartley coal, above quoted, on the authority of Mr. Longridge, the composition of the sample under trial was as follows:

Carbon ---- --81.5 per cent.
Hydrogen--- 5.2 per cent.
Nitrogen -- 1.5 per cent.
Oxygen -- 6.2 per cent.
Sulphur --- 1.1 per cent.
Ash -- 4.5 per cent.

100.0

The quantity of air chemically consumed in the combustion of one pound of this fuel by formula (1), is 144 cubic feet at 62°.   The actual quantity of air that was admitted below and about the fire was, according to Mr. Longridge, 158 cubic feet, being 14 cubic feet, or 9¾ per cent, in excess, when smoke was entirely prevented. He mentions, at the same time, that with the ordinary system of stoking, when dense smoke was given off, the quantity of air that passed through the furnace, exclusively through the grate, was only at the rate of 100 cubic feet per pound of coal.   This was little more than equal to what was sufficient to burn the fixed portion of the coal.

Below we give a table giving the conditions and results of perfect combustion for the fuels in common use.

As this table is from English sources the heat of combustion and equivalent evaporative power of the coal is much higher than our American coals warrant, as the tables of American coals, p. 20 will show.   In applying the table to practical cases, the surplus air which reduces the efficiency must be taken into account.   It is good practice to get in actual evaporation 60 per cent. of the theoretical evaporative power for the poorer, and 70 per cent. for the better kinds of coal:

TABLE NO. 9.

Total Heat Evolved by Various Fuels and their Equivalent Evaporative Power, with the Weight of Oxygen and Volume of Air Chemically Consumed.

D. K. C.

FUEL.	Weight of Oxygen Consumed per lb. of Fuel.	Quantity of Air Consumed per lb. of Fuel.		Total Heat of Combustion of 1 lb. of Fuel.	Equivalent Evaporative Power of 1 lb. of Fuel from and at 212°.
	Lbs.	Lbs.	Cu. ft. at 62°.	B. T. U.	Lbs.
Hydrogen	8.0	34.8	457	62000	62.40
Carbon --- { Making Carbonic Acid.	2.66	11.6	152	14500	15.0
Sulphur	1.00	4.35	57	4000	4.17
Coal, average dessicated.	2.45	10.7	140	14700	15.22
Coke, " "	2.49	10.81	142	13548	14.02
Lignite, perfect..........	2.04	8.85	116	13108	13.57
Asphalt	2.74	11.85	156	17040	17.64
Wood, dessicated.......	1.40	6.09	80	10974	11.36
" 25 per cent. moisture	1.05	4.57	60	7951	8.20
Straw, 15% per ct. moist.	0.98	4.26	56	8144	8.43
Petroleum	3.20	14.33	188	20411	21.13
Petroleum Oils...........	4.12	17.93	235	27531	28.50

For average American coals the following table gives good approximate results for the temperature and volume of gases, *in the furnace*, under the varying conditions of practice. In applying it the actual quantities of air used should be measured by an anemometer:

TABLE NO. 10.

Temperature of Combustion and Volumes of Products.

J. M. W.

TEMPERATURE OF GAS, FAHRENHEIT.	Supply of Air in lbs. per lb. of Fuel.		
	12 lbs.	18 lbs.	24 lbs.
	Volume of Air or Gases in Cubic Feet at Each Temperature.		
32	150	225	300
68	161	241	322
104	172	258	344
212	205	307	409
392	259	389	519
572	314	471	628
752	369	553	738
1112	479	718	957
1472	588	882	1176
1832	697	1046	1395
2500	906	1359	1812
3275	1136	1704	--------
4640	1551	--------	--------

Brown Palace Hotel, Denver, Colo.
Heat and Power from 520 H. P. of Heine Boilers.

COAL.

Coal is by far the most important fuel in use. The cases where wood is used are exceptional, and becoming more so as population increases and timber becomes scarce and more in demand for structural purposes. Very favorable local conditions are necessary before fuel oils or gases can compete with coal. It is interesting to trace the gradual increase in the demand for coal.

In England coal was first used in the twelfth century, and was then and long after known as sea-coal to distinguish it from char-coal. This name was given it from the fact that it was first believed to be a marine product, being gathered among the seaweed and other wreckage cast up by the waves on Northumbrian beaches. Later on the name was given to coal brought from over the sea.

About the year 1200 the English began to dig coal systematically for the use of their smiths and lime burners. In 1281 the entire coal trade of Newcastle on Tyne amounted to about $500 a year. In 1307 the brewers, dyers, etc., of London had so generally adopted coal in their works that a commission to abate the smoke nuisance was instituted. Its powers and methods were far less restricted than those of similar commissions now being very generally instituted in American cities.

In dwellings coal was not used till the middle of the fourteenth century, since chimneys had first to be invented, but early in the fifteenth century we find Falstaff sitting "at the round table, by a sea-coal fire."

In 1577 a writer says in regard to the coal mines, "Theyr greatest trade beginneth now to grow from the forge into the kitchin and hall." When the Stuarts came to the English throne they made the use of coal fashionable, so that in 1612 a writer states that it had become "the generale fuell of this Britaine Island." "Coking" coal (originally "cooking" it) came in vogue about 1640, and in 1656 an English knight anticipated the St. Louis Smoke Committee of 1892 in attempting to introduce coke for domestic purposes. But as late as 1686 sea-coal and pit-coal were considered "not useful to metals," and char-coal still held the field in smelting furnaces. But during the next fifty years, lead, tin and finally iron furnaces began to use coal. Soon after the gradual development of steam power began. In 1800 the total production of coal in Great Britain had reached ten million tons. In 1891 the records show 185,479,126 tons of which about 1-6 was exported, 1-6 was for domestic use, and the other 2-3 was consumed in the arts and manufactures.

In the United States up to 1860 the use of wood as fuel, for dwellings, for factories, steamboats and locomotives was quite general, except in the anthracite coal districts. But since then the use of bituminous coal has increased rapidly and steadily for all purposes.

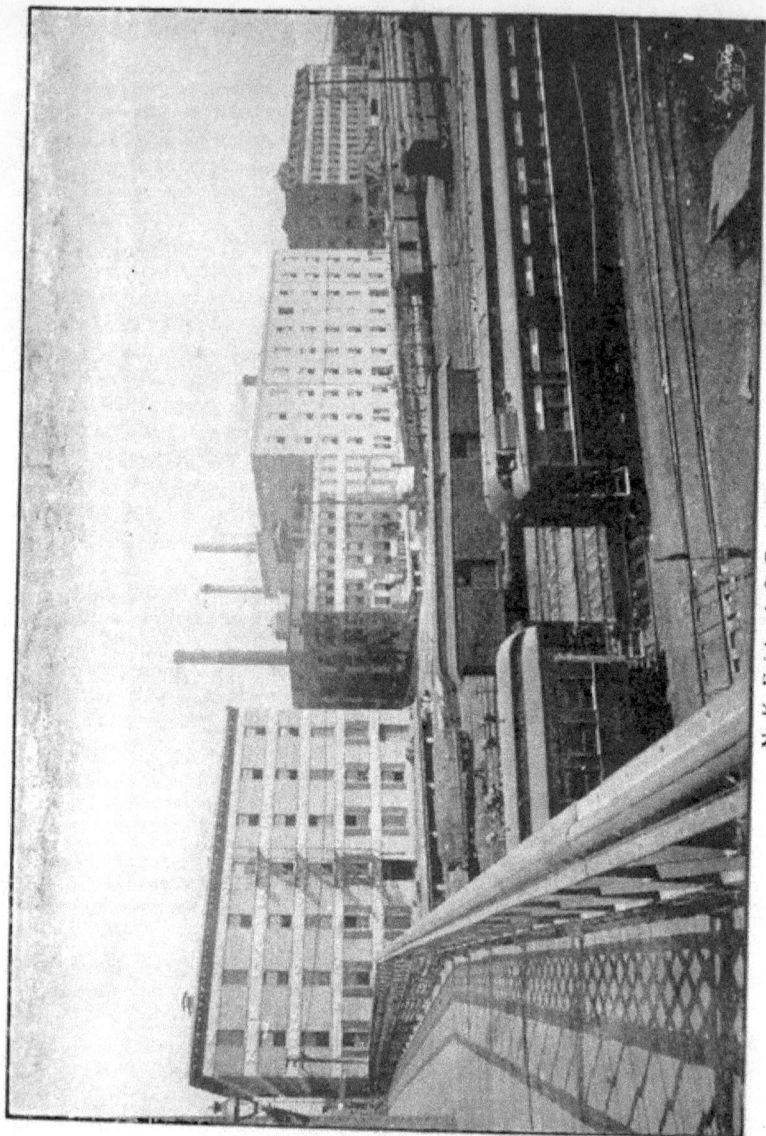

N. K. Fairbank & Co.'s Factories,
CHICAGO, ILL.

The following table gives the amounts of coal produced during the last twelve years :

Amount of Coal, in Tons of 2000 lbs., Mined in the United States Since 1880.

YEAR.	ANTHRACITE.	ALL OTHERS.	TOTAL.
1880	26,249,711	47,398,286	73,647,997
1881	31,920,018	56,327,412	88,247,430
1882	32,614,507	65,588,241	98,202,748
1883	35,418,353	72,663,765	108,082,118
1884	36,558,478	73,836,730	110,395,208
1885	38,335,973	74,273,838	112,609,811
1886	39,035,446	75,624,846	114,660,292
1887	42,088,196	88,887,109	130,975,305
1888	46,619,564	98,850,642	145,470,206
1889	39,656,635	98,460,065	138,116,702
1890	46,468,640	109,604,971	156,073,611
1891	50,665,431	118,878,517	169,543,948
1892	49,735,744	122,033,611	171,769,355
1893	47,354,563	128,823,364	176,177,927
1894	52,010,433	117,950,348	169,960,781
1895	51,785,122	135,118,193	186,903,315
1896	48,010,616	137,640,276	185,650,892

In the United States a long ton of coal is 2240 lbs.

In the United States a short ton of coal is 2000 lbs.

In Illinois, Kentucky and Missouri 80 lbs. of bituminous coal make a bushel.

In Pennsylvania, 76 lbs. of bituminous coal make a bushel.

In Indiana 70 lbs. of bituminous coal make a bushel.

A cubic foot of solid anthracite coal weighs 93.5 lbs.

Forty-two cubic feet of prepared anthracite coal weigh one long ton.

COAL may be arranged in five classes :

1st.   Anthracite, or blind coal, consisting almost entirely of free carbon.

2d.   Dry bituminous coal, having from 70 to 80 per cent. of carbon.

3d.   Bituminous caking coal, having from 50 to 60 per cent. of carbon.

4th.   Long flaming or cannel coal, having from 70 to 85 per cent. of carbon.

5th.   Lignite, or brown coal, containing from 56 to 76 per cent. of carbon.

In the United States the anthracites are found mainly in the eastern portion of the Allegheny Mountains and the Rocky Mountains of Colorado ; the dry bituminous coals in Maryland and Virginia ; the caking coals in the great Mississippi Valley ; the cannel coals in Pennsylvania, Indiana and Missouri ; the lignites in Colorado, Texas and Washington. The second and third classes furnish the best steam coal.

The following table, compiled from a number of analyses of coals bought in the open market may prove of value, bearing in mind what we said of the difference between theoretical and practical heating powers. (See p. 15.)

We will add what a noted German engineer, Mr. F. Bode, says on this point: "*The calculation of the calorific value of a given coal from an elementary analysis is unreliable, and often gives results greatly at variance with an actual calorimetric test.*"

Table of American Coals.

Heating and Evaporative Power

COAL. Name or Locality.	B. T. U. per Pound.	Theoretical Evap. in lbs. water from and at 212°.	COAL. Name or Locality.	B. T. U. per Pound.	Theoretical Evap. in lbs. water from and at 212°.
ARKANSAS.			**IOWA.**		
Coal Hill, Johnson Co	11812	12.22	Milwaukee Pea	10240	10.60
Huntington Co	11757	12.16	Thornburgh	10690	11.07
Huntington Co	11906	12.32	Muchikinock	11370	11.77
Huntington Co	12537	12.97	Good Cheer	8702	9.01
ILLINOIS.			**KENTUCKY.**		
			Kanawah	12689	13.13
Big Muddy, Jackson Co	11466	11.87	Kanawah	13345	13.81
Big Muddy, Jackson Co	11529	11.93			
Big Muddy, Jackson Co	11781	12.19	**MARYLAND.**		
Big Muddy, Jackson Co	11200	11.60	George's Creek Cumberland	13700	14.18
Carterville	11481	11.89	George's Creek Cumberland	13400	13.87
Carterville	12383	12.71	George's Creek Cumberland	12800	13.25
Carterville	11498	11.90			
Carterville	11407	11.81	**MISSOURI.**		
Carterville	11337	11.73			
Carterville	11700	12.12	Bevier	9890	10.24
Carterville	11400	11.80	Cannel	11832	12.24
Colchester	9848	10.19	Carter	10880	11.26
Colchester Slack	9035	9.35	Elston	12656	13.82
Collinsville, Madison Co	10143	10.50	Freeburg	11436	11.83
Dumferline Slack	9401	9.73	Henry	10466	10.83
Duquoin, Jupiter	10710	11.08	Keene	10956	11.34
Glen Carbon	9675	10.01	K. T	10448	10.81
Glen Carbon	9804	10.14	Lump	9414	9.75
Gillespie, Macoupin Co	9739	10.09			
Girard, Macoupin Co	9954	10.30	**NEW MEXICO.**		
Girard, Macoupin Co	10269	10.63	Coal	11756	12.17
Heitz Bluff, St. Clair Co	10332	10.69			
Heitz Bluff, St. Clair Co	10576	10.95	**OHIO.**		
Hurricane	11868	12.28	Hocking Valley	13309	13.78
Muddy Valley	11718	12.14	Jackson Co	12343	12.77
Oakland, St. Clair Co	10395	10.76	Jackson Co	11600	12.01
Paradise	11340	11.73			
St. Bernard	10080	10.44	**PENNSYLVANIA.**		
St. Clair	9261	9.58			
St. Clair	10294	10.65	Clearfield	14000	14.49
St. Clair	10647	11.02	Pittsburgh	13104	13.46
St. John, Perry Co	9765	10.10	Pittsburgh Gas	13035	13.49
St. John, Perry Co	9828	10.18	Pittsburgh Slack	11739	12.15
Streator, LaSalle Co	11403	11.80	Reynoldsville	12981	13.44
Trenton, Clinton Co	10584	10.96	Wilkesbarre	13563	14.04
Trenton, Clinton Co	11245	11.63	Youghiogheny	12936	13.39
Turkey Hill	11255	11.64	Youghiogheny	12600	13.03
Turkey Hill	11260	11.65	Youghiogheny	13480	13.95
Vulcan	9450	9.78	Youghiogheny	13287	13.75
Vulcan	10626	11.00	Youghiogheny	12909	13.36
			Youghiogheny	13222	13.69
INDIANA.			Youghiogheny	12278	12.71
			Youghiogheny	13305	13.77
Block	10407	10.77	Youghiogheny	12600	13.04
			Youghiogheny	13111	13.47
INDIAN TERRITORY.			Youghiogheny	12487	12.92
			Youghiogheny	12600	13.04
Atoka	11088	11.47	Youghiogheny	13309	13.77
Choctaw Nation	12789	13.23	Youghiogheny	13158	13.60
McAllister	13287	13.75	Oil (Crude)	17268	17.88
McAllister	12800	13.25	Oil (Crude)	16801	17.39

Table of American Coals—Continued.

COAL. Name or Locality.	B. T. U. per Pound.	Theoretical Evap. in lbs. water from and at 212°.	COAL. Name or Locality.	B. T. U. per Pound.	Theoretical Evap. in lbs. water from and at 212°.
TENNESSEE.			**WASHINGTON.**		
Glen Mary, Scott Co	13167	13.63	Carbon Hill	12316	12.75
Lump	12600	13.04	Carbon Hill	12085	12.51
Lump	12215	12.65	Carbon Hill	12866	13.32
TEXAS.			**WEST VIRGINIA.**		
Ft. Worth	9450	9.78	New River	13374	13.84
Ft. Worth	11803	12.22	New River	12806	13.26
VIRGINIA.			New River	12800	13.25
			New River	12962	13.52
Pocohontas	13363	13.83			
Pocohontas	13029	13.49			

The average proximate analysis of a few of the commonest coals are given in the following table:

	Moisture.	Volatile Matter.	Fixed Carbon.	Ash.	Sulphur.
Ordinary Illinois	9.90	33.40	43.80	12.80	3.35
Best Illinois	6.40	30.60	54.60	8.30	1.78
Pennsylvania Bituminous	1.70	31.80	60.10	6.40	.84
Pennsylvania Anthracite	2.00	6.40	78.40	13.20	
New River, W. Va	.85	18.40	77.60	2.90	0.26

Boiler Plant of the Orleans Street Ry. Co.,
NEW ORLEANS, LA.
500 H. P. Heine Boilers.

As foreign results in the work of both boilers and engines are frequently brought to our notice by the professional press, it will be convenient to have some tables of English, French and other foreign coals, for purposes of comparison, and they are here given:

TABLE No. 13.

Average Composition of British and Foreign Coals, with their Weight, Bulk, Heat of Combustion and Evaporative Power.

D. K. C.

COAL. Averaged Groups.	Specific Gravity.	WEIGHT AND BULK. 1 Cu. ft. Solid. Lbs.	1 Cu. ft. Heaped. Lbs.	Bulk of 1 Ton Heaped. Cu. ft.	COMPOSITION. Carbon. Per ct.	Hydr'n. Per ct.	Nitro. Per ct.	Sulph. Per ct.	Oxygen. Per ct.	Ash. Per ct.	Coke Produced from Coal. Per ct.	Total Heat of Combustion of 1 lb. Units of Heat. Units.	Equivalent Evap. from and at 212°. Lbs.	Evaporative Power of 1 lb. Coal from and at 212°. By trial. Lbs.
Welsh	1.315	82.0	53.1	42.7	83.78	4.79	0.98	1.43	4.15	4.91	73	15123	15.66	9.05
Newcastle	1.256	78.3	49.8	45.3	82.12	5.31	1.35	1.24	5.69	3.77	61	15203	15.74	8.01
Derbyshire and Yorkshire	1.292	80.6	47.2	47.4	79.68	4.94	1.41	1.01	10.28	2.65	59	14616	15.13	7.58
Lancashire	1.273	79.4	49.7	45.2	77.90	5.32	1.30	1.44	9.53	4.88	58	14602	15.12	7.94
Scotch	1.260	78.6	50.0	42.0	78.53	5.61	1.00	1.11	9.69	4.03	54	14868	15.39	7.70
Average	1.279	79.8	50.0	44.5	80.40	5.19	1.21	1.25	7.87	4.05	61	14876	15.40	8.13
Anthracite, Ireland	1.590	99.6	62.8	35.7	80.03	2.30	0.23	6.76	INCLUDED IN ASH.	10.80	90	13031	13.49	9.85
Patent Fuels	1.167	73.6	65.2	34.4	83.40	4.97	1.08	1.26	2.79	5.93	74.2	15176	15.71	9.20
FOREIGN: Van Dieman's Land					65.80	3.50	1.30	1.10	5.58	22.71		11713	12.13	
Chili					63.56	5.43	0.82	2.50	14.84	13.31		12571	13.01	
Lignite, Trinidad					65.20	4.25	1.33	0.69	21.69	6.84		12091	12.52	

Steel Framing of the Carnegie Building,
PITTSBURGH, PA.,
Completed building contains 505 H. P. Heine Boilers.

Composition and Heating Power of French Coals.
D. K. C.

COAL.	Carbon.	Hydrogen.	Oxygen.	Nitrogen.	Water.	Ash.	Fixed Carbon.	Volatile Elements	Observed.	Calculated.	Theoretical Evaporative Power.
	Per Cent.	Per Cent.	Per Cent.	Per Cent.	Per Cent.	Per Cent.	Per Cent.	Per Cent.	B. T. U.	B. T. U.	Pounds of Water.
RONCHAMP.											
No. 1	76.5	4.4	3.0	1.1	.....	15.0	61.7	23.3	14357	13820	14.86
No. 2	68.6	4.0	4.7	1.1	0.8	20.8	55.6	23.6	13743	12430	14.23
No. 3	76.2	4.1	5.9	1.0	.....	12.8	62.3	24.9	14085	13590	14.59
No. 4	73.1	3.8	4.9	1.0	.....	16.2	62.4	21.4	13995	12960	14.49
Average	73.6	4.1	4.6	1.5	.....	16.2	60.5	23.3	14045	13220	14.54
SARREBRUCK.*											
Dudweiler	71.3	4.1	9.2	0.5	1.8	13.1	53.5	33.4	13833	12880	14.32
Altenwald	69.3	4.3	9.9	0.5	2.5	13.5	52.9	33.6	13320	12720	13.79
Heinitz	70.3	4.3	11.5	0.5	1.8	11.6	53.7	34.7	13548	12860	14.03
Friedrichsthal	67.8	4.2	13.8	0.5	1.0	12.7	50.2	37.1	13647	12440	14.13
Louisenthal	64.7	3.9	15.0	0.5	3.6	12.3	47.3	40.4	12665	11800	13.11
Sulzbach	73.3	4.6	9.6	0.5	1.6	10.4	.....	.....	13608	13480	14.09
Von der Heyt	70.6	4.5	11.2	0.5	2.7	10.5	.....	.....	13865	13030	14.36
BLANZY.											
Montceau	66.1	4.4	13.2	0.5	10.3	5.0	.....	.....	12720	12300	13.17
Anthracitic	67.0	3.6	5.9	0.5	21.0	2.0	.....	.....	12825	11950	13.28
Creuzot, Anthracite	87.4	3.5	3.2	0.5	3.6	1.8	.....	.....	16108	14850	16.68

Combustion of Coal.

"When coal is exposed to heat in a furnace, a portion of the carbon and hydrogen, associated in various chemical unions, as hydro-carbons, are volatilized and passed off. At the lowest temperature, naphthaline, resins, and fluids with high boiling points are disengaged; next, at a higher temperature, volatile fluids are disengaged; and still higher, olefiant gas, followed by common gas, light carburetted hydrogen, which continues to be given off after the coal has reached a low red heat. What remains after the distillatory process is over, is coke, which is the fixed or solid carbon of coal, with earthy matter, the ash of the coal.

Taking the fixed carbon, or coke remaining in the furnace after the volatile elements are distilled off, for round numbers at 60 per cent., the following is an approximate summary of the condition of the elements of average coal, after having been decomposed, and prior to entering into combustion:

100 POUNDS OF AVERAGE COAL IN THE FURNACE.

COMPOSITION.	LBS.	DECOMPOSITION.	LBS.
Carbon { Fixed	60	Fixed Carbon	60
Carbon { Volatilized	20	Hydrocarbons	24
Hydrogen	5	Sulphur	1 1-4
Sulphur	1 1-4	Water or Steam	9
Oxygen	8	Nitrogen	1 1-5
Nitrogen	1 1-5	Ash	4
Ash	4		
About	100		100

showing a total useful combustible of 86¼ per cent. of which 26¼ per cent. is volatilized. While the decomposition proceeds, combustion proceeds, and the 26¼ per cent. of volatilized portions, and the 60 per cent. of fixed carbon, successively, are burned.

* These are now German Coals.

The sulphur and a portion of the nitrogen are disengaged in combination with hydrogen, as sulphuretted hydrogen and ammonia. But these compounds are small in quantity, and, for the sake of simplicity, they have not been indicated in the above synopsis.

There are three modes of supplying coal to ordinary furnaces by hand firing, namely : spreading, alternate, and coking firing. In spreading firing the charge of coal is scattered evenly over the whole surface of the grate, commencing generally at the bridge, and working forward to the door. In alternate firing the charge of coal is laid evenly along half the width of the grate at a time, from back to front, each side alternately. In coking firing the charge of coal is thrown on to the dead plate and the front part of the bars and left there for a time, in order that the mass may become coked through, and when that is done the mass is pushed back towards the bridge, and another charge is thrown on to the front of the fire in its place. In this way the gases are gradually evolved from the coal at the front, while a bright fire is maintained at the back.

It is thought advantageous, in slowly burning furnaces having long flues, that the fuel should be slightly moist, and that the ash pits should be supplied with water, from which steam may be generated by the heat radiated downwards from the fire, and passed through the firegrate. The access of water to the fuel lessens the "glow fire" or flameless incandescence of the fixed carbon on the grate, and increases the quantity of flame by forming carbonic oxide and hydrogen gases in its decomposition into its elements, oxygen and hydrogen, and the reduction, by the oxygen, of the carbonic acid already formed in the furnace. The newly made gases are afterwards burned in the flues. The presence of moisture, even in coke, gives rise to flame in the flues, and reduces the intensity of the heat in the glow fire. The combustion, in fact, is deferred, or distributed ; and it is on this principle that moist bituminous coals are most effective in furnaces having long flues, as in Cornish boilers.

That two coals of identical composition may possess very different heating powers is evidenced by comparing the bituminous coals of Creuzot and Ronchamp, which have the following nearly identical compositions, reckoning the coal as dry and pure, or free from ash :

	Carbon. Per cent.	Hydrogen. Per cent.	Oxygen. Per cent.	Heating Power.
Creuzot	88.48	4.41	7.11	17320
Ronchamp	88.32	4.78	6.89	16339

while there is a difference of six per cent. in the actual heating powers. Correspondingly, the Creuzot coal had only 19.6 per cent. of volatile matters, while the Ronchamp coal yielded 27 per cent.

Lignite and Asphalt.

Brown lignite is sometimes of a woody texture, sometimes earthy. Black lignite is either of a woody texture, or it is homogeneous, with a resinous fracture. Some lignites, more fully developed, are of a schistose character, with pyrites in their composition. The coke produced from various lignites is either pulverulent, like that of anthracite, or it retains the forms of the original fibres. Lignite is less dense than coal.

Asphalt, like lignite, has a large proportion of hydrogen. It has less than 9 per cent. of oxygen and nitrogen, and thus leaves 8¼ per cent. of free hydrogen, and it accordingly yields a porous coke.

The average composition of perfect lignite and of asphalt may be taken in whole numbers as follows:

	Lignite.	Asphalt.
Carbon	69 per cent.	79 per cent.
Hydrogen	5 "	9 "
Oxygen and Nitrogen	20 "	9 "
Ash	6 "	3 "
	100	100
Coke, by laboratory analysis	47 "	9 "

The lignites are distinguished from coal by the large proportion of oxygen in their composition—from 13 to 29 per cent.

The heating powers of lignite and asphalt are respectively measured by 13,108 units, and 17,040 units.

WOOD.

Wood, as a combustible, is divisible into two classes: 1st. The hard, compact, and comparatively heavy woods, as oak, beech, elm, ash; 2d. The light-colored, soft, and comparatively light woods, as pine, birch, poplar.

In the forests of Central Europe, wood cut down in winter holds, at the end of the following summer, more than 40 per cent. of water. Wood kept for several years in a dry place retains from 15 to 20 per cent. of water. Wood which has been thoroughly desiccated will, when exposed to air under ordinary circumstances, absorb 5 per cent of water in the first three days; and will continue to absorb it until it reaches from 14 to 16 per cent., as a normal standard. The amount fluctuates above and below this standard, according to the state of the atmosphere. Ordinary firewood contains, by analysis, from 27 to 80 per cent. of hygrometric moisture.

The woods of various trees are nearly identical in chemical composition, which is practically as follows, showing the composition of perfectly dry wood, and of ordinary firewood holding hygroscopic moisture:

TABLE NO. 15.

	Desiccated Wood.	Ordinary Firewood.
Carbon	50 per cent	37.5 per cent.
Hydrogen	6 per cent	4.5 per cent.
Oxygen	41 per cent	30.75 per cent.
Nitrogen	1 per cent	0.75 per cent.
Ash	2 per cent	1.5 per cent.
	100 per cent.	75.0 per cent.
Hygrometric water		25.0 per cent.
		100.0

The quantity of intersticial space in a closely packed pile of wood, consisting of round uncloven stems, is 30 per cent. of the gross bulk; for cloven stems, the intersticial space amounts to from 40 to 50 per cent.

English oak—a hard wood—weighs 58 lbs. per solid cubic foot; its specific gravity is .93. Yellow pine—a soft wood—weighs 41 lbs. per solid cubic foot; its specific gravity is .66.

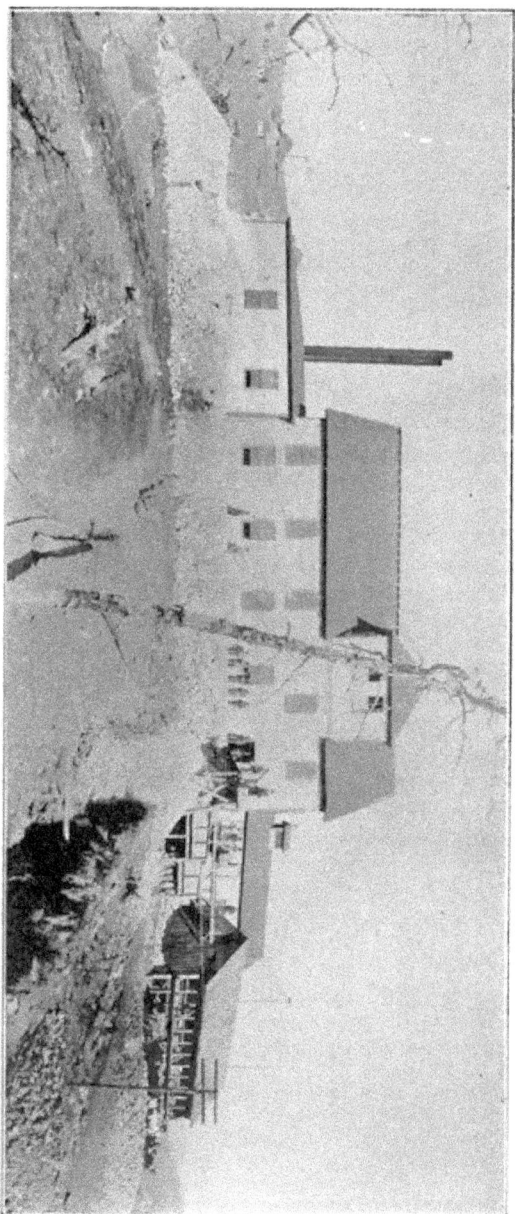

Plant of the Independence Mine,
VICTOR, COLO.
Contains 900 H. P. Heine Boilers.

A cord of pine wood—that is, of pine wood cut up and piled—in the United States, measures 4 feet by 4 feet by 8 feet, and has a volume of 128 cubic feet. Its weight in ordinary condition averages 2700 lbs.; or 21 lbs. per cubic foot.

The quantity of air chemically consumed in the complete combustion of one pound of perfectly dry wood, by rule 1, page 13, is 80 cubic feet at 62° F., or 6.09 lbs. of air. The quantity of burnt gases for 1 lb. of perfectly dry wood are

TABLE No. 16.

	By Weight.		By Volume.	
	Lbs.	Per cent.	Cu. ft. at 62°F.	Per cent.
Carbonic acid	1.83	21.7	15.75	14.4
Steam	0.54	6.4	11.40	10.4
Nitrogen	6.08	71.9	82.01	75.2
Totals	8.45	100.0	109.16	100.0

showing that there are 8½ lbs., or 109 cubic feet, at 62° F., of burnt gases per pound of wood, 13 cubic feet to the pound.

The total heat of combustion of perfectly dry wood, by rule 4, page 14, is 10974 units, which is 75 per cent. of that of coal, and is equivalent, by rule 5, to the evaporation of 11.36 lbs. of water from and at 212° F.

When the wood holds 25 per cent. of water, there is only 75 per cent. or three-quarter pound of wood substance in one pound; and the total heat of combustion is 75 per cent. of 10974 units, or 8230 units, which is only 56½ per cent. of that of average coal. Similarly, the equivalent evaporative power is reduced to 8.52 lbs. of water from and at 212°, of which the equivalent of a quarter of a pound is appropriated to the vaporizing of the contained moisture—that is to say, for evaporating one-quarter pound of water, supplied at 62° F., the quantity of heat is 1116°÷4=279 units, and the net available heat for service is 8230—279=7951 units per pound of fuel holding 25 per cent. of water.*

In order to obtain the maximum heating power from wood as fuel, it is the practice, in some works on the continent of Europe,—as glass works and porcelain works,—where intensity of heat is required, to dry the wood fuel thoroughly, even using stoves for the purpose, before using it."

D. K. C.

The American Society of Mechanical Engineers in their Rules for Boiler Tests allow 1 lb. of wood = 0.4 lb. of coal; or 2½ lbs. of wood = 1 lb. of coal. Other authorities estimate 2¼ lbs. of dry wood = 1 lb. of good coal. One pound of wood is practically equivalent to one pound of any other kind of wood *equally dry*.

TABLE NO. 17.

1 cord of hickory or hard maple weighs .... 4500 lbs. and = 2000 lbs. coal.
1 cord of white oak weighs .... 3850 lbs. and = 1711 lbs. coal.
1 cord of beech, red oak, or black oak weighs ... 3250 lbs. and = 1445 lbs. coal.
1 cord of poplar, chestnut, or elm weighs .... 2350 lbs. and = 1044 lbs. coal.
1 cord of average pine weighs .... 2000 lbs. and = 890 lbs. coal.

* This figure may be used for a close approximation in comparing a certain kind of wood to a known coal. Suppose the calculated heat in a pound of the coal to be 13025 B. T. U., and an actual boiler test showed an evaporation of seven pounds of water per pound of coal. Then 13025:7951::7:4.28, *i. e.*, you may expect to evaporate about 4.28 lbs. of water per pound of the wood in the same boiler.

In substituting any kind of wood for coal under a boiler, the dimensions of the furnace must be increased, preferably mainly in the height, so that by carrying a greater depth of fuel nearly as much by weight may be present in the furnace as was usual or necessary with the coal.

" BAGASSE."

Bagasse is the fibrous portion of the cane left after the juice has been extracted from it in the mill. There is a great difference in the chemical composition of bagasse ; that from tropical canes shows a greater proportion of combustibles.

Prof. L. A. Becnel, in an address to the Louisana Sugar Chemists Association, said: " The judicious use of bagasse as fuel is perhaps one of the most important questions with which we have to deal, and which has a direct bearing on the reduction of cost of manufacture." He then quotes from Mr. N. Lubbock that 4.83 lbs. of bagasse from a double mill making 72 per cent extraction, or 5.98 lbs. of single mill bagasse of 66 per cent extraction, will produce about as much heat as one pound of Scotch coal.

Mr. L. Metesser, as the result of a number of tests in Cuba and Mexico, reports from 4.25 to 5 lbs. of 70 per cent bagasse as equal to one pound of good coal.

Tropical cane and the bagasse remaining after mill extraction are of about the following composition :

	Cane.	66% Bagasse.	70% Bagasse.	72% Bagasse.
Woody Fibre	12.5	37	40	45
Water	73.4	53	50	46
Combustible Salts	14.1	10	10	9
	100 lbs.	100 lbs.	100 lbs.	100 lbs.

Taking these figures as a basis, and remembering that the water in the bagasse has to be first brought up from an average temperature of say 86° F., to steam under atmospheric pressure, requiring 1060 H. U., and that this steam has to be raised to the average stack temperature say 300° higher, and taking the specific heat of gaseous steam at 0.475, which would give say 142 H. U. more, therefore a total of 1200 H. U. per pound of water. Mr. Lubbock found 51 per cent of carbon in the woody fibre, and 42.1 per cent of carbon in the combustible salts. Since a pound of carbon in perfect combustion will liberate 14500 H. U., we will have in 100 lbs. of 66 per cent bagasse, 334660 H. U., from which we must deduct 63600 H. U. as absorbed by the water, leaving 271060 H. U. available as fuel. In like manner, we have in the 72 per cent bagasse 387730 H. U., from which we must deduct 55200 absorbed by the water, leaving 332530 H. U. available.

In comparing this with good Youghiogheny coal of say 13000 H. U., and good Scotch coal of 14800 H. U. calorific value, we find the fuel value of the 66 per cent bagasse to be :

    5   lbs. bagasse equals one pound Youghiogheny coal,
    5.52 "   "   "   "   "   Scotch coal,

and that of the 72 per cent bagasse to be

    3.85 lbs. bagasse equals one pound Youghiogheny coal,
    4.35 "   "   "   "   "   Scotch coal.

It will probably require considerably more of the Louisiana bagasse than of the tropical bagasse, since it has about 25 per cent less woody fibre than the latter.

Mr. Becnel, estimates with 75 per cent Louisiana bagasse as a basis, that " To manufacture one ton of cane into sugar and molasses, it will take from 145 to 190 lbs. additional coal by the open evaporation process; from 85 to 112 lbs. additional coal with a double effect," and with triple effect it appears the bagasse alone would do the work, and have enough steam to spare to run engines, grind cane, etc. " If this has not yet been accomplished in Louisiana, may it not be due more to imperfect boiler and evaporating plants than to a deficiency in heat producing properties of the bagasse?"

The above of course can only be taken as approximately correct. The results will vary greatly according to the kind of boilers and furnaces used. From the nature of this fuel, it follows that it should be fed continuously into a very hot fire brick chamber, and that plenty of room must be left in the furnace and boiler setting to accommodate the large volume of gas and steam produced by the bagasse.

Bridge Mill Power Co..
PAWTUCKET, R. I.
Contains 610 H. P. of Heine Boilers.

The higher the per cent of extraction the more fuel value the bagasse will have, and as it will necessarily contain less moisture, the larger proportion of this enhanced fuel value becomes available in the boiler furnace. The improvement in boiler plants will thus naturally go hand in hand with improved methods of extraction.

TAN AND STRAW.

Tan.

"Tan, or oak bark, after having been used in the process of tanning, is burned as fuel. The spent tan consists of the fibrous portion of the bark. According to M. Peclet, five parts of oak bark produce four parts of dry tan; and the heating power of perfectly dry tan, containing 15 per cent of ash, is 6100 English units, while that of tan in an ordinary state of dryness, containing 30 per cent of water, is only 4284 English units. The weight of water evaporated at 212° by one pound of tan, equivalent to these heating powers, is as follows:

	Perfectly Dry.	With 30% of Moisture.
Water supplied at 62°	5.46 lbs.	3.84 lbs.
Water supplied at 212°	6.31 lbs.	4.44 lbs.

(See note under Wood.)

Straw.

The composition of straw, in its ordinary air-dried condition, is given by Mr. John Head, as follows:

TABLE NO. 18.

	Wheat Straw, per cent.	Barley Straw, per cent.	Mean, per cent.
Carbon	35.86	36.27	36.
Hydrogen	5.01	5.07	5.
Oxygen	37.68	38.26	38.
Nitrogen	.45	.40	.425
Ash	5.00	4.50	4.75
Water	16.00	15.50	15.75
	100.00	100.00	100.00

The weight of pressed straw is from 6 lbs. to 8 lbs. per cubic foot.

Heat of Combustion of Straw.

For straw of mean composition, the total heat generated is, by rule 4, equal to 145 [36 + (4.28×5)] = 8323 units of heat, or the evaporation of 7.46 lbs. of water from and at 212° F. Deducting the heat absorbed in evaporating the constituent water, 15¾ per cent, or .16 lb., equal to 1116×.16= 179 units, the available heat is 8323—179 = 8144 units, equivalent to the evaporation of 7.30 lbs. of water from and at 212°.

(See note under Wood.)

LIQUID FUELS.

Petroleum is a hydrocarbon liquid which is found in abundance in America and Europe. According to the analysis of M. Sainte-Claire Deville, the composition of fifteen petroleums from different sources was found to be practically the same. The average specific gravity was .870. The extreme and the average elementary compositions were as follows:

Chemical Composition of Petroleum.

Carbon--------82.0 to 87.1 per cent. Average, 84.7 per cent.
Hydrogen-------11.2 to 14.8 per cent. Average, 13.1 per cent.
Oxygen ------- 0.5 to 5.7 per cent. Average, 2.2 per cent.

100.0

The total heating and evaporative powers of one pound of petroleum having this average composition are, by rules 4 and 5, as follows :

Total heating power-------- = 145 [84.7 + (4.28×13.1)] = 20411 units.
Evaporative power: evaporating at 212°, water supplied at 62° = 18.29 lbs.
Evaporative power: evaporating at 212°, water supplied at 212° = 21.13 lbs.

Petroleum oils are obtained in great variety by distillation from petroleum. They are compounds of carbon and hydrogen, ranging from $C_{10} H_{24}$ to $C_{32} H_{64}$; or, in weight ;

TABLE NO. 20.

Chemical Composition of Petroleum Oils.

		Mean.
From $\begin{cases} 71.42 \text{ Carbon} \\ 28.58 \text{ Hydrogen} \end{cases}$ to	$\begin{cases} 73.77 \text{ Carbon} \text{----------} \\ 26.23 \text{ Hydrogen} \text{--------} \end{cases}$	72.60 27.40
100.00	100.00	100.00

The specific gravity ranges from .628 to .792. The boiling point ranges from 86° to 495° F. The total heating power ranges from 28087 to 26975 units of heat; equivalent to the evaporation, at 212°, of from 25.17 to 24.17 lbs. of water supplied at 62°, or from 29.08 lbs. to 27.92 lbs. of water supplied at 212°.

D. K. C.

Oil as a Fuel.

"Inasmuch as the use of oil as a fuel is now engaging the serious attention of many of our principal engineers and manufacturers, we beg leave to submit for your consideration the following advantages which are claimed for oil as against coke, coal or wood as a fuel.

1st. A petroleum fire can be held in perfect control by one man of ordinary intelligence, by the mere turning of a valve. He can increase or decrease the fire at will, and can maintain steam or heat at any desired point. When the fire is properly regulated to produce the heat required, it can be kept at that point with but slight attention, so slight, indeed, that one man can fire and care for a battery of from eight to ten 100-horse-power boilers without difficulty. By turning a valve you can instantly extinguish the fire, if occasion does not require its continuous use, and it can be again started with almost the same rapidity with a few shavings or sticks of wood. There is no waste, as with coal, when the work is done.

2d. The heat generated with petroleum fire is much more uniform than that produced with coal or wood. The fire is not as sensitive to the fluctuation of the weather as other fires. A great advantage is gained in running machinery where regularity of speed is desirable. A constant supply of steam may be furnished, and there is no reduction of steam pressure in consequence of the replenishing of fires.

Toronto Municipal and County Buildings,
TORONTO, ONT., CANADA.
Contains 700 H. P. Heine Boilers.

3a. Economy of Boiler Capacity.— It has been demonstrated that one pound of oil will evaporate the water of more than two pounds of coal. The heat units of crude petroleum have been erroneously stated to be 27531. The fact is, that the correct figure, 20240 heat units, has been repeatedly arrived at of late, after many tests with the best instruments to be obtained for that purpose. In comparing the calorific properties of petroleum it must be borne in mind that with coal there is an enormous waste of matter, such as sulphur, slate and earthy substances which are practically incombustible, and which do not add in the generation of heat. While coal theoretically contains about 14300 heat units, that figure is, by reason of these impurities, reduced to about 8000.   Pure carbon — charcoal, for instance — contains 14500 heat units.   Considering, therefore, the imperceptible waste in the burning of oil, and the excessive waste in the burning of coal, the conclusion is reached that while theoretically the relative proportion of heat evolved in the combustion of oil compared with coal is as 20.2 is to 14.3, the proportion practically considered, is in favor of oil as 19 is to 8, or 8.5 at the furthest. We may quite safely assume, then, that the heating capacity of oil is considerably more than twice that of coal as far as now shown. With a clean boiler, properly attended, and with the best of coal fuel, well stoked night and day — with every care to insure combustion and to avoid waste, the evaporation obtained in some isolated cases specially recorded has been as high as $9\frac{1}{2}$ pounds.   In our every day experience, however, we find that eighty out of a hundred boilers will not vaporize more than from 7 to $7\frac{1}{2}$ pounds of water per pound of fuel. On the other hand, oil tests, which, while sufficiently conclusive for the present, have not, by any means, been carried to the furthest limit, show the evaporation from 17.56 to 18.5 pounds of water per pound of oil consumed, from and at 212° Fahrenheit.

4th.   Economy in labor, cleanliness and safety are secured, as in burning oil complete combustion may be obtained.   There is no shoveling of ashes, and consequently there is a great saving in labor.   The absence of sparks and cinders and the ability to extinguish the fire instantly in case of danger, makes it very desirable when considered with a view to safety.

5th.   There being no necessity for opening doors for the introduction of fuel, there is no fluctuation of heat, and no sudden chilling of the flues and boiler.   The absence of sulphur in the fuel makes its action on the metal of the boiler and the flues much less destructive than coal, and the flues remain cleaner and in better condition to absorb the heat.

6th.   Oil or Residuum, is without doubt, the coming fuel on locomotives and ocean steamers, and by its use a great annoyance to passengers in the emission of cinders and smoke will not only be entirely avoided, but less than one-half the room formerly used for coal will be required to store the oil for fuel, and only one-third the weight will be carried, thus saving a great deal of room in storage, which will enable ship owners to carry an additional quantity of freight, or to increase speed to the same amount of power. Besides this, where 70 stokers are now required to unload coal on ocean steamers, at least 60 could be dispensed with, and the work be done without the labor of shoveling coal and the great discomfort from heat and dust.

7th.   Regarding the proper construction of furnaces for the consumption of oil, it may be said that there is no occasion for having the combustion chamber as large as when burning coal.   The latter article, being solid

The Largest Steam Dredge Ever Built—A. D. 1895.

U. S. Mississippi River Dredge Boat "Beta,"
1,500 H. P. Heine Safety Boilers.

3,100 H. P. Double and Triple Compound
Condensing Engine.

Maximum guaranteed work in excavating sand,
2,400 cubic yards per hour.

Maximum actually obtained on test, 7,700
cubic yards per hour.

matter, requires more time for decomposition, and the elimination of the products and supporters of combustion. Coal fuel requires a large fire chamber and the means for the introduction of air beneath the grate-bars to aid combustion. Compared with oil, the combustion of coal is tardy and requires some aid by way of a strong draft. Oil having no ash or refuse, when properly burned, requires much less space for combustion for the reason that, being a liquid, and the compound of gases that are highly inflammable when united in proper proportions, it gives off heat with the utmost rapidity, and at the point of ignition is all ready for consumption. The changes required to burn oil in a coal furnace may be made at a nominal cost, so that even in this respect no additional expense is necessary for a change for the better.

8th. Three barrels of oil, each of 42 gallons, equal and slightly exceed the heating capacity of one ton of coal. The oil weighs 913.5 pounds, and may be purchased and delivered in tank cars at any point in the United States at a very low figure. It should be remembered that oil need not be shoveled from the cars to the furnace, it needs no stoking nor poking, it leaves neither cinders nor ashes to be carted away, and it makes no smoke. With an oil furnace, one man may attend to a dozen boilers without any further assistance.

9th. The fact of being able to produce with oil a perfectly clear, white fire, free from ashes, smoke, dust and soot, which can be kept under control and regulated to any degree of heat required, makes its use invaluable in electric plants, in the manufacture of glass, steel, crockery, stoneware, sewer pipe, brick, lime, and in fact almost any business where such a fire is required."

C. W. O.

In November, 1894, the Baldwin Locomotive Works, of Philadelphia, equipped an engine for burning fuel oil and obtained the results stated below:

TESTS OF OIL FUEL ON LOCOMOTIVE.

DATE, 1894.	No. 1. November 13.	No. 2. November 18.	No. 3. November 25.
Weight of train, approximate, lbs	1,308,160	1,216,120	1,480,640
Number of cars	25 and 20	30	26
Length of run, miles	89.7	54.9	52.3
	h. m. s.	h. m. s.	h. m. s.
Time of run	6 27 00	2 56 41	3 20 0
Running time	5 14 48	2 23 26	2 48 9
Average steam pressure, lbs		171	170
Oil consumption, total lbs	6,637	3,200.7	3,703
Total gallons	905		
	lbs.	lbs.	lbs.
Per hour	1,003.2	1,086.9	1,110.9
Per square foot of grate	237	114.32	132.25
Per square foot of grate per hour	38.3	38.82	39.68
Per square foot of heating surface	3.13		
Per square foot of heating surface per hour,	0.49		
Water evaporated: total lbs	70,933	34,151.7	39,169.2
Total from and at 212° F	85,622	41,465.1	46,291.6
Per hour	10,998		
Per hour from and at 212° F	13,280	14,082.2	13,887.5
Per lb. of oil	10.69	10.67	10.58
Per lb. of oil from and at 212° F.*	12.90	12.95	12.50
Per square foot of heating surface	33.47	16.12	18.48
Per square foot of heating surface per hour,	5 19	5.48	5.54
Per square foot of heating surface per hour from and at 212° F		6.64	6.55

*Without deducting the steam consumed for vaporizing the oil, or the entrainment.

The report on the experiments points out that oil has several advantages over coal: 1, no smoke if the firing is properly done; 2, no sparks; 3, no terminal labor in cleaning fires, hauling away ashes and loading coal, which labor is said to amount sometimes to 50 cents per ton of coal consumed; 4, the engine is always ready for service; 5, the fire is always clean and there is no danger of its being torn up by a heavy exhaust or by the engine slipping. Tests of the oil used showed 84 gravity, 140 flash and 190 fire. In conclusion, it is stated that to determine the value of oil, it is necessary to know the evaporative power of the boiler for each pound of fuel burned, which depends greatly upon the ratio of heating surface to grate area, and the volume consumed in a given time. These conditions do not seem to affect the consumption of oil, the evaporation being about the same per pound of oil for all rates of combustion, it being impossible to consume the oil without a proper supply of air, and, as no smoke is made, no unconsumed fuel escapes from the smokestack, as is the case with soft coal. The following formula is given for obtaining the value of oil, as compared with coal, as a locomotive fuel, the result being the price per gallon at which oil will be the equivalent of coal. In this formula the cost of both oil and coal must be the cost delivered on the engine, and not the purchasing price:

$$A = \frac{B \times 10.7 \times 7}{2,000 \times C}$$

A = price per gallon at which oil will be equivalent of coal; B = cost of coal per ton, plus the cost of handling (say 50 cents per ton); C = evaporative power of coal.

From a lecture at the Naval War College, Newport, R. I., delivered by P. A. Engineer John R. Edwards, U. S. N., in August, 1895, we quote the following:

With reference to the use of liquid fuel on locomotives, it is interesting to refer to the results obtained in England by Mr. James Holden, Locomotive Superintendent of the Great Eastern Railway, by the process invented and adopted by him. On the locomotive using liquid fuel there is an absence of constant and laborious firing; the requisite pressure of steam is easily obtained by an almost imperceptible movement of the injector valve; there is an absence of smoke, and a great uniformity of pressure.

In the inaugural address of the President of the Society of Engineers, in February, 1894, he gave a description of these locomotives and their working cost. He stated that an express engine using 35.4 pounds of coal per mile, consumed under similar circumstances 11.8 pounds of coal and 10.5 pounds of liquid fuel, or a total of 22.3 pounds of fuel.

The advantages of the Holden system are summed up as follows:

1st. With an ordinary grate, steam can be easily raised without working the injector.

2d. Fuel can be interchanged according to the state of the market.

3d. With a thin coal fire, oil can be shut off at will without running the risk of chilling the fire box.

4th. When standing, the coal fire will maintain steam.

For several years a number of locomotive engines on the Great Eastern Railway have used liquid fuel, and one of these engines is recorded to have traveled 47,000 miles without a single failure or accident. But the great difficulty in extending the use of liquid fuel in England is the impossibility of obtaining a sufficient supply at a low cost, otherwise it would be very generally used, considering the great calorific effect and the practical advantages of its application.

It has been very recently stated that since the introduction in the naval ships of liquid fuel, the cost in Italy has increased one hundred and fifty per cent (150 per cent).

One of the highest officials of the Pennsylvania Railroad asserted that the great cost attending its use was a bar to its introduction in the locomotives of that road.

On the other hand, there are some places where it can be secured more cheaply than coal.

The question of cost, therefore, depends upon location.

A great writer upon this subject has said: "We must look for the best results from petroleum, both economically and technically, in those uses where the improved product of the manufactured article more than counterbalances the difference in price of the two kinds of fuel."

CHIEF ENGINEER SOLIANI'S MONOGRAPH ON LIQUID FUEL.

Undoubtedly one of the best articles that has been published on this subject is the paper of Chief Engineer Soliani of the Italian Navy, which was read at the International Engineering Congress. He starts in with the various kinds of petroleum used, gives the chemical composition, what its actual calorific value as fuel is, and then goes on to tell about the experiments in Russia, where it was first used on vessels in the Volga region and on the Caspian Sea. He then gives us the pulverizing process adopted by Mr. Urquhardt, and then brings us down to to-day's actual modern experience in the Italian Navy.

A careful study of this paper shows:

1st. That the only form of liquid fuel which is absolutely safe for use on board of ship is what is known as petroleum refuse, which is a thick viscous fluid of about the consistency of tar or very thick molasses. This has to be sprayed or pulverized, either by jets of air or steam, for use in the furnaces.

2d. The pulverizers form the principal element in the whole arrangement for burning liquid fuel, and many kinds have been used or tried, or simply patented. The Russian pulverizers are all worked by steam, and they appear to be the best, because a pulverizer using steam may be worked well with air, or any other suitable gaseous fluid with little or no alteration.

3d. Where pulverizers are not used a compressor for forcing the air is employed. Its great weight and space occupied forms a very serious objection to the compressor.

4th. That the use of liquid fuel by the Russians is almost confined to the Volga region and the Caspian Sea. There the wood is scarce and costly and also very bulky. Coal is extremely expensive. One very remarkable fact in connection with the use of liquid fuel on Russian vessels is that the difficulty with marine boilers of making up the waste of steam entailed by the pulverizers does not exist for the steamers running along the Volga River. It is lessened, in case of the sea steamers, by the fact that the great bulk of the Caspian trade is from Baku and other ports south to Astrakhan, where fresh water is available in abundance, and can be stored by the steamers both for outward and homeward passages.

5th. Italy, on account of its position and of its deficiency of coal, was naturally interested in the matter. And that country, which even our naval experts have, in years past, mistakenly reported as having adopted this fuel for its war vessels, confines the practice to a few torpedo boats. For their large vessels they do not contemplate the regular use of liquid fuel. Pulverizers, however, are fitted in order that they may be held in readiness for the same object as forced draft.

6th. The system of mixing petroleum spray with the coal seems to be on the increase in the French and Italian navies, and furnishes a ready means of rapidly increasing the steam pressure and speed, above that of the natural draft.

7th. That the measure of success in the burning of liquid fuel will depend upon the efficiency and durability of the pulverizer. Less than three years ago the Italians believed that they had solved this question for naval purposes by the invention of the Curriberti atomizer. They are now rather doubtful about this sprinkler satisfying all their wants. The French, who are following them more closely than any other nation, are about to use their own pulverizers.

There is no one who has made a more protracted and scientific investigation of its capabilities than Mr. Isherwood, and this is the result of his observations on liquid fuel as a combustible for naval purposes. In summarizing the work of the Experimental Board, of which he was president, he writes:

"The experiments in question embrace those made with Col. Foote's apparatus at the Charleston Navy Yard, and those made with other apparatus on different boilers in the New York Navy Yard, all of them, I believe, of considerable value, but never reported in full with the exception of one made about ten years ago, and which is now on the files of the Bureau of Steam Engineering. In every case the patentees abandoned the trials before they were completed, owing to the failure of their apparatus.

"The liquid oil has, in all cases, to be transformed into oil gas before it can be burned. This transformation can be made by the direct application externally of heat to the liquid, but the temperature of the oil on the vaporizing surface is higher than the temperature required to decompose it, the result being deposition of solid carbon in the form of coke which soon fills the vaporizing vessels and renders them useless. This coke is frequently so hard that cold chisels can scarcely detach it, and if thrown into a fire even in small fragments, it burns with excessive slowness, like graphite. Whenever the vaporizing vessel is subjected to a high temperature like that of a boiler furnace, the decomposition of the oil and deposition of coke go rapidly on, so that in the course of a few hours any vessel of practical size is filled by it. All apparatus exposed to anything like furnace or flame temperature will inevitably fail from these causes in the future as they have in the past. To make trials with such devices will

Flour Mills of The C. C. Washburn's Flouring Mills Co.,
MINNEAPOLIS, MINN.
Contains 2700 H. P. Heine Boilers.

merely serve to confirm this fact. The smaller the vaporization vessel, and the higher the temperature to which it is exposed, the more quickly will it fail.''

INSTALLATION OF THE SYSTEM AT THE CHICAGO EXPOSITION.

At the World's Fair at Chicago the boilers which furnished the steam for driving the machinery were all fed with crude oil. The conditions there were, of course, quite different from what would prevail aboard ship, but they were all in favor of a more successful burning of the liquid fuel. Lake Michigan with its supply of fresh water was near. There was no question of either weights or space occupied to be taken into consideration. The seven representative boiler firms which were pitted against each other sent excellent men to look out for their respective plants. The piping was arranged in the most efficient manner, it not being necessary to make extra bends or angles in order that it would clear a hatch or opening, as might occur on board ship. And yet an official report says, ''The quantity of petroleum used for firing the main boiler plant at the World's Columbian Exposition amounted to upwards of 31,000 tons, and the work done is stated to have totalled 32,316,000 horse power hours. This makes the consumption of oil about 2.1 pounds per horse power hour.'' This report would tend to dispose of some of the claims of the thermal efficiency of liquid fuel. Is it possible that the commercial article is not so rich in hydrogen as that furnished for experimental purposes?

From time to time we hear of the success attained by one of the Italian cruisers on a short run with this fuel. A careful investigation invariably shows, that when oil was used in connection with coal, the speed over that of natural draft was increased. There is not one single instance on record where the burning of liquid fuel, either alone or in combination with coal, developed the speed of horse power that was secured with coal under forced draft.

For the past year, the Austrians have been experimenting with it. It is said that for every pound of residuum they were able to burn, seven-tenths of a pound of water in the form of steam was required to spray it. They have not yet been convinced of its merits for naval purposes, for not a single boat in the Austrian service has yet been fitted permanently with atomizers for burning this fuel.

A careful reading of the professional papers in regard to the success of the French with this combustible furnishes one with such information as the following: ''The question is altogether in a state of tentative experiment, and the fuel will have to be tried in different boilers and under severe conditions before adoption in large vessels.'' Of another vessel it is written: ''The results are said to be good, but not definite.'' Concerning three torpedo boats it is written: ''The experiments have been more or less successful.''

Capt. A. M. Hunt, formerly of the U. S. N., read an exhaustive paper before the Technical Society of the Pacific Coast, October 5th, 1894, on the results obtained from oil at the Mid-Winter Fair, from which the following extracts are taken:

A certain amount of eye training is necessary to judge whether or not the oil is being burned so as to give the maximum heating effect. With proper manipulation of the burner, it being of proper design, and a careful regulation of the air, an almost flameless combustion can be obtained. The furnace should never be filled with an opaque, luminous flame, although many so-called practical oil men claim that such a combustion will give the best results as regards evaporation.

The best results at the Fair were always obtained by so manipulating the burner, with the air full on, as to get a blue, Bunsen-burner-like flame, and then shutting off steam and air until a tinge of luminosity began to show, chasing through the furnace in waves. Under such conditions the carbon in the oil is being entirely consumed, and the air supply is being limited just to the point necessary for its consumption. Luminosity indicates the presence of unconsumed carbon, and consequent failure to obtain the full heating effect. After the furnace once becomes thoroughly heated there should be absolutely no evidence of smoke issuing from the chimney.

The flame must not be allowed to impinge directly against the iron of a boiler. Overheating of the metal is apt to be the result. If a solid particle or drop of the oil strikes the comparatively cold metal the volatile matter is driven off and a carbon deposit left, which, becoming incandescent, and being in direct contact with the iron, burns it.

At the Mid-Winter Fair there was a chance to determine the efficiency of oil as fuel, and the results of several of the tests conducted by Mr. E. C. Meier, assistant in charge of the boiler plant, and the author, are appended. In these tests both the feed water and the oil were measured by Worthington meters, and while meter tests are always regarded with great suspicion by engineers, these tests were so conducted as to be quite reliable.

Before any tests were made the feed water meter was thoroughly calibrated by weighing the water passed through it. At first it was found impossible to get concordant results from different sets of weighings, especially when the temperature of the feed water was high. A pressure gauge was finally placed on the boiler side of the meter, and the discharge valve passing water into the weighing barrels, set so as to maintain on the meter the ordinary boiler pressure, and the pump run at a speed which furnished feed water at the rate required for the boilers under test. It was found, after so doing, that the results of different sets of calibrations did not vary more than two-tenths of a pound per cubic foot registered. This was as close as the limit of accuracy of the scales.

The oil meter was calibrated in the same manner. This meter was provided with small vents to enable any gas which might collect at the top of meter chambers to be removed. Great care was taken to avoid any chance for errors. The blow lines were blanked, and feed valves so arranged that no feed water could pass into boilers not in use.

The thermometers and steam gauges used were corrected by standards. The only tests for dryness of the steam were made with an ordinary barrel calorimeter, such as Thurston describes, and the results showed the steam to be practically dry.

The average result of all the tests made was 14.2 pounds of water evaporated from and at 212 degrees Fahrenheit per pound of oil, and the highest evaporation of any test is 15.13 pounds to one.

It is a very common thing to hear oil men say that they have obtained an evaporation of 17 and even 18 pounds of water per pound of oil, and you will find recorded the results of tests showing such evaporations, attested by all the details of columns of figures and calculations. There is suspicion that the oil used in such tests must have been thinned down with liquified hydrogen, or that the experimenter deceived himself, willfully or otherwise.

The theoretical evaporation of oil is about 20.7 pounds to one. An evaporation of 18 to one would indicate a boiler efficiency of about 87 per cent, and assuming a furnace temperature of 2,400 degrees Fahrenheit, the temperature of the issuing gases would be 312 degrees. If the temperature of the issuing gases was 450 degrees, as would be more probable, the furnace temperature would be 3,461 degrees, rather too high for comfort.

The results of certain tests made by the Edison Light and Power Company, of San Francisco, Cal., were as follows:

Evaporation with California oil			13.1	pounds	to 1
"	"	Peru oil	12.1	"	to 1
"	"	White Ash coal	6.68	"	to 1

The California oil used weighed 320 pounds to the barrel. The Peru oil used weighed 294 pounds to the barrel.

1 pound of California oil = 1.96 pounds of coal.
1 pound of Peru oil = 1.81 pounds of coal.

Accepting the results at the Mid-Winter Fair, an evaporation of 15 to 1 can be obtained with oil, using the Heine boilers.

In January, 1895, a series of tests were made with crude oil on Heine boilers at the Harrison Street station of the Chicago Edison Company, with the following average results:

Size of boiler	500
Lbs. of water per lb. oil from and at 212°	14.57
H. P. developed	593
Per cent. over rating	18.6

For fuel purposes two kinds of oil are used, crude petroleum, usually from Lima, O., and residuum after distilling off the lighter oils.

The Lima crude petroleum oil comes to this (St. Louis) market in tank cars holding 6000 gallons. The price is 1.8 cents per gallon, to which must be added $5.00 per car for switching, etc. Even under favorable conditions, therefore, as to location of boiler plant, the cost of this oil delivered to the boiler will be at least 2 cents per gallon. A gallon of this oil weighs 6.9 lbs. The theoretical heat value of this oil is about 20,000 heats units, equivalent to a theoretical evaporation of 20.7 lbs. of water. Assuming an efficiency of 80 per cent., the evaporation in practice would be 16.56 lbs. of water per pound of oil. The cost of evaporating 1000 lbs. of water would therefore be 17.54 cents. With a bituminous coal giving an evaporation in practice of 5 lbs. of water per pound of coal and costing $1 25 per ton, the same work could be done for 12.5 cents, a difference in favor of the coal of 40.32 per cent. It will be observed also that the conditions assumed in this calculation are especially favorable to the oil.

The fuel oil or residuum weighs about 7.3 lbs. per gallon, and has a calorific power of 16,880, or a theoretical evaporation of 17.47 lbs. water per pound of oil. At 3 cents a gallon and under the conditions assumed above the cost of evaporating 1000 lbs. of water would be 29.28 cents or 134 per cent. more than when using the coal.   W. B. P.

From these varying statements it is clear that local conditions must decide when liquid fuel can be used to advantage.—F. i. at the World's Columbian Exposition at Chicago, more than 25,000 H. P. of boilers are being run by fuel oil piped there from Lima, O. Aside from the saving in dust, noise, soot and ashes—made apparent by the white uniforms of the stokers and the white boiler fronts—it would be an impossibility to bring in coal and carry off ashes for this huge plant without seriously interfering with the passenger traffic. It would require eighty cars for coal and twenty for ashes daily.   E. D. M.

300 H. P. Heine Boiler being moved.

FUEL GAS.

Gaseous Fuel has so many apparent practical advantages over any other form of fuel, that it may be properly regarded as the ideal fuel. Near Pittsburgh, and in some favored districts of Indiana, *Natural Gas* has been found in such quantities that—for some years at least—immense manufacturing industries have been based on it. Manufacturers who have once realized its advantages are loth to surrender them and would gladly welcome some kind of artificial gas to take its place—if this can be made cheap enough to compete with the local coal. Inventors have been prolific of processes and devices to fill this demand.

As there are certain fixed and well defined conditions on which the fuel value of such gases depends, we give below extracts from papers on the subject by well known experts, which will enable the careful engineer to estimate in each particular case pretty closely whether gas may be economically substituted for coal.

Mr. Emerson McMillin, in October, 1887, made an exhaustive investigation of the subject of fuel gas from which we extract the following :

" The relative calorific value of the various gases now in use for heating and for illumination have been frequently published, yet, in the discussion of this subject we cannot well avoid a reproduction of some of the figures.

" Notwithstanding the fact that tables of this character have been so often published, we are all more or less confused occasionally by seeing statements made that make the comparison totally different from our preconceived ideas as to their relative calorific values.

" This confusion occurs from the fact that at one time we see the comparison of the gases made by weight, and at another time the comparison is made by volume. We present here the comparison made both by weight and by volume, and shall use natural gas as the unit of value in both comparisons :

TABLE NO. 21.

Relative Values.

	By Weight.	By Volume.
Natural gas	1,000	1,000
Coal gas	949	666
Water gas	292	292
Producer gas	76.5	130

" The water gas rated in the above table—as you will understand—is the gas obtained in the decomposition of steam by incandescent carbon, and does not attempt to fix the calorific value of illuminating water gas, which may be carbureted so as to exceed, when compared by volume, the value of coal gas.

TABLE NO. 22.

Composition of Gases.

	VOLUME.			
	Natural Gas.	Coal Gas.	Water Gas.	Producer Gas.
Hydrogen	2.18	46.00	45.00	6.00
Marsh gas	92.60	40.00	2.00	3.00
Carbonic oxide	0.50	6.00	45.00	23.50
Olefiant gas	0.31	4.00	0.00	0.00
Carbonic acid	0.26	0.50	4.00	1.50
Nitrogen	3.61	1.50	2.00	65.00
Oxygen	0.34	0 50	0.50	0.00
Water vapor	0.00	1.50	1.50	1.00
Sulphydric acid	0.20	- - - -	- - - -	- - - -
	100 00	100.00	100.00	100.00

Philadelphia & Reading R. R. Station,
PHILADELPHIA, PA.
Contains 425 H. P. Heine Boilers.

Composition of Gases.

	Natural Gas	Coal Gas.	Water Gas.	Producer Gas.
		WEIGHT.		
Hydrogen	0.268	8.21	5.431	0.458
Marsh gas	90.383	57.20	1.931	1.831
Carbonic oxide	0.857	15.02	76.041	25.095
Olefiant gas	0.531	10.01	0.000	0.000
Carbonic acid	0.700	1.97	10.622	2.517
Nitrogen	6.178	3.75	3.380	69.413
Oxygen	0.666	1.43	0.965	0.000
Water vapor	0 000	2.41	1.630	0.686
Sulphydric acid	0.417	.....	.....	.....
	100.000	100.000	100.000	100.000

" Some explanations of these analyses are necessary. The natural gas is that of Findlay, O. The coal gas is probably an average sample of coal gas, purified for use as an illuminant. The water gas is that of a sample of gas made for heating, and consequently not purified, hence the larger per cent. of CO_2 that it contains.

"Since calculating the tables used in this paper, I am satisfied that the sample of water gas is not an average one. The CO is too high, and H is too low. Were proper corrections made in this respect, it would increase the value in heat units of a pound, but not materially change the value when volume is considered, and as that is the way in which gases are sold, the tables will not be recalculated.

"The producer gas is that of an average sample of the Pennsylvania Steel Works, made from anthracite, and is not of so high grade as would be that made from soft coal.

"The natural gas excels, as shown in Table 21, because of the large per cent. of marsh gas. In no other form, in the gases mentioned, do we get so much hydrogen in a given volume of gas.

" It is the large per cent. of hydrogen in the coal gas that makes it so nearly equivalent to the natural gas in a given weight, but much of the hydrogen in coal gas being free, makes it fall far short of natural gas in calorific value per unit of volume.

"A further comparison of the value of the several gases named may be made by showing the quantity of water that would be evaporated by 1000 feet of each kind of gas, allowing an excess of 20 per cent. of air, and permitting the resultant gases to escape at a temperature of 500 degrees. This sort of comparison probably has more practical value than either of the others that have been previously given. We will assume that the air for combustion is entering at a temperature of 60 degrees.

Water Evaporation.

	Natural Gas.	Coal Gas.	Water Gas.	Producer Gas
Cubic feet gas	1000	1000	1000	1000
Pounds water	893	591	262	115

"The theoretical temperature that may be produced by these several gases does not differ greatly as between the three first-named. The producer gas falls about 25 per cent. below the others, giving a temperature of only 3441° F.

"Water gas leads in this respect, with a temperature of 4850°.

"A comparison of the resultant products of combustion also shows water gas to possess merit over either natural or coal gas, when the combustion of equal quantities—say 1000 feet—is considered. An excess of 20 per cent. of air is calculated in the following table :

TABLE NO. 25.
Resultant Gases of Combustion.

Quantity—1000 ft.	Natural Gas.	Coal Gas	Water Gas.	Producer Gas.
Weight of gas before combustion, lbs.	45.60	32.00	45.60	77.50
Steam	94.25	69.718	25.104	6.92
Carbonic acid	119.59	68.586	61.754	36.45
Sulphuric acid	0.36	....	....	....
Nitrogen	664.96	427.222	170.958	126.57
Total weight after combustion	879.16	565.526	257.816	169.94
Pounds oxygen for combination	167.46	107.961	43.149	19.67

"You will observe, by the following table, that, with the exception of producer gas, each kind gives off nearly one pound of waste gases for each pound of water evaporated. This quantity includes 20 per cent. excess of air:

TABLE NO. 26.
Weights of Water Evaporated and Resultant Gases.

	Natural Gas.	Coal Gas.	Water Gas.	Producer Gas.
Weight of water evaporated	893.25	591.000	262.000	115.100
Weight of gases after combustion	879.16	565.526	257.816	169.945

"The vitiation of the atmosphere per unit of value in water evaporation is practically the same in water gas as in natural gas.

"However, the excess of oxygen does no harm, and the steam and nitrogen can not be regarded as very objectional products. The gas that robs the air permanently of the most oxygen, and produces the greatest quantity of carbonic acid per unit of work, must be classed as the most objectionable from a sanitary standpoint.

TABLE NO. 27.
Oxygen Absorbed and Carbonic Acid Produced.

In Combustion.	Natural Gas.	Coal Gas.	Water Gas.	Producer Gas.
Pounds of oxygen absorbed per 100 lbs. water evaporated	18.75	18.27	16.47	17.96
Pounds of CO_2 produced per 100 lbs. water evaporated	13.40	11.60	23.57	31.70
Oxygen absorbed plus CO_2 produced	32.15	29.87	40.04	49.66

"Here, then, it is shown that if pollution by carbonic acid and the impoverishment by the absorption of oxygen are equally deleterious to the atmosphere, coal gas stands at the head as being the least objectionable."

Mr. McMillin then goes into an elaborate calculation of a mixture of gases, which would combine the good qualities of the three artificial gases compared, which he finds to be "in per cent., coal gas 20.35, water gas 32.17, producer gas 47.48."

After calculating the cost of such gas, he proceeds:

"Here we may note some features, that to my mind are interesting; that is, the cost of various gases per 1,000,000 units of heat which they are theoretically capable of producing.

"In working out these figures I put wages, repairs and incidentals and the cost of the ton of good gas coal at $2.00, and a ton of hard coal or coke at the same price, and the quantities of production as follows: Coal gas from soft coal, 10,000 feet ; water gas from hard coal, 40,000 feet; and producer gas, 150,000 feet.

TABLE NO. 28.

Cost per 1,000,000 Units of Heat.

Coal gas ....734,976 units, costing 20.00 cents = 27.21 cents per mill.
Water gas.......322,346 units, costing 10.88 cents = 33.75 cents per mill.
Producer gas ...117,000 units, costing 2.58 cents = 22.05 cents per mill.
Our mixture....323,115 units, costing 7.88 cents = 24.39 cents per mill.

"Thus it will be seen that after all coal gas costs but 11.6 per cent. more per unit of heat than the mixture that we have worked out, while water gas, per unit of heat, costs 38.38 per cent. more than the mixed product."

After a discussion of methods of delivery and the various uses for the fuel gas, he concludes:

"The demand for fuel gas, like the demand for electric light, has come to stay. It will not down. Scientific investigators, as well as the public, insist that there ought to be, and must be, a change in the mode of domestic and industrial heating. Our present systems are not in keeping with the progress of the nineteenth century."

Professor D. S. Jacobus—Oct., 1888—says: "It is proposed to give an estimate of the cost at which carbureted and uncarbureted water gas will have to be sold in order to compete successfully for steam boiler use with anthracite coal.

"The following are analyses of the gas direct from the generator, and of the same after it has been carbureted for illuminating purposes:

TABLE NO. 29.

Analyses of Water Gas.

| | Per cent by Volumes. | |
	I. Uncarbureted.	II. Carbureted.
Nitrogen	4.69	2.5
Carbonic acid	3.47	.3
Ethylene	.0	12.5
Oxygen	.0	.2
Benzole vapor	.0	1.5
Carbonic oxide	36.80	29.0
Marsh gas	2.16	24.0
Hydrogen	52.88	30.0
	100.00	100.00

"Experiments were made to verify the ratio between the heats of combustion of the gas before and after being carbureted, by determining the time and quantity of gas required to evaporate a given weight of water contained in an open vessel, heating by means of a gas stove. The burner of the stove used for burning the carbureted gas caused air to be mingled with the gas before the latter was burned, thus producing a colorless flame. For the uncarbureted gas, the burner was of the Argand pattern, and the air was not mingled with the gas before it was burned. The results of the tests are given in the following table:

TABLE NO. 30.

GAS	Weight of water evaporated.	Temperature of water at start.	Temperature of room in which experiments were made.	Pressure of gas at meter in inches of water.	Time required to heat water to its boiling point.	Time required to heat and completely evaporate the 4 lbs. of water.	Number of cubic feet of gas burned.	Amount of water that would be evaporated from and at 212° F. by the amount of gas burned.	
								Actual.	Per 1000 cu. ft. gas.
	Lbs.	Deg. F.	Deg. F.	Inches.	Min.	Min.	Cu. ft.	Lbs.	Lbs.
Uncarbureted	4	56	61	4	12	79½	36.1	4.64	128.5
Uncarbureted	4	59	62	4	11¾	79½	34.8	4.64	133.3
Carbureted	4	56	61	5½	8½	69	19.4	4.64	239.2
Carbureted	4	59	62	5½	10	73¼	18.2	4.64	254.9
Carbureted	4	68	66	4½	8¼	63	19.8	4.60	232.3

"From this table we obtain an average evaporation, per 1000 cubic feet of uncarbureted gas, of 130.9 lbs. of water from the temperature of 212° and at atmospheric pressure, and for the carbureted gas an average of 242.1 lbs. That is, the experiments tend to show that the ratio of the calorific power of the uncarbureted to that of the carbureted gas, by volume, is $\frac{130.9}{242.1} = 0.54$.

"The volume of one pound of air at 62° F. is 13.14 cubic feet, hence the calorific power of the gases per cubic foot should be:

$$\text{Uncarbureted } \frac{7373 \times 0.505}{13.14} = 283.3 \text{ H. U.}$$

$$\text{Carbureted } \frac{13317 \times 0.632}{13.14} = 640.5 \text{ H. U.}$$

"One pound of good Lackawanna coal has a calorific power of 13000 heat units. The number of cubic feet of gas required to produce the same heating effect as is produced in burning one ton of coal will therefore be:

$$\text{Uncarbureted gas } \frac{13000 \times 2000}{283.3} = 91780.$$

$$\text{Carbureted gas } \frac{13000 \times 2000}{640.5} = 40590.$$

"The efficiency of a boiler will be about the same when burning gas for fuel as when coal is used. (?) If, therefore, we disregard the difference in the cost of attendance, in favor of using gaseous fuel, 40590 cubic feet of carbureted, or 91780 cubic feet of uncarbureted gas, must be sold for the same price as one (short) ton of coal, in order to successfully compete with the latter for use under steam boilers.

600 H. P. Plant of Heine Boilers in Power House of People's Ry. Co.,

ST. LOUIS, MO.

" We will estimate the probable excess of cost involved in firing with coal above that of gas-firing. In burning natural gas, it has been found that one man is able to attend to boilers aggregating 1500 horse-power, whereas, if coal is used, 200 horse-power will be a fair figure. It therefore requires, in the case of large plants, only about $\frac{2}{15}$ as much labor to fire with gas as is involved in firing with coal. If the wages of firemen are \$2.50 per day of twelve hours, the cost of firing one ton of coal in boilers that require four pounds of coal per boiler horse-power per hour, will be

$$\frac{2.5 \times 2000}{12 \times 200 \times 4} \text{ dollars} = 52.1 \text{ cents.}$$

The cost of firing with gas would be $\frac{2}{15}$ of this amount so that the extra cost of firing with coal would equal $\frac{13}{15}$ times 52.1 equal 45 cents. To make a liberal allowance in favor of the gaseous fuel we will assume that the extra labor of firing each ton of coal costs 50 cents.

" The following table contains the prices at which the gas must be sold per 1000 cubic feet in order to be as cheap as coal for boiler use; first, disregarding the difference in the cost of attendance, and secondly, assuming that the extra cost of firing with coal is 50 cents per ton.

TABLE No. 31.

Cost of Burning Water Gas.

Cost of Coal Delivered to Boiler Room. Per Ton of 2000 Lbs.	PRICE OF GAS PER 1000 CUBIC FEET.			
	Not Including Difference in Cost of Attendance.		Including Difference in Cost of Attendance.	
	Carbureted.	Uncarbureted.	Carbureted.	Uncarbureted.
\$6.00	14.8 cts.	6.5 cts.	16.0 cts.	7.1 cts.
5.00	12.3 cts.	5.4 cts.	13.6 cts.	6.0 cts.
4.00	9.9 cts.	4.4 cts.	11.1 cts.	4.9 cts.
3.00	7.4 cts.	3.3 cts.	8.6 cts.	3.8 cts.
2.00	4.9 cts.	2.2 cts.	6.2 cts.	2.7 cts.
1.00	2.5 cts.	1.1 cts.	3.7 cts.	1.6 cts.

"As the cost of manufacturing the uncarbureted gas, by the present processes, not including interest on capital invested, is from 10 to 20 cents per 1000 cubic fèet, it does not appear possible to employ it economically for a boiler fuel, until considerable reduction in cost of production shall have been made."

Professor Wm. B. Potter, March, 1892, says:

" The convenience and economy attending the use of natural gas in a number of localities in this country have led many people to believe that fuel gas, made from coal at large central stations, and distributed to factories and works, is the fuel of the future which will not only clear all chimneys but reduce all fuel bills as well. While it is unquestionably true that fuel gas is especially adapted for household use and will play an important part in the future for such use, it is equally true that as a fuel for raising steam it can never compete in the matter of economy with coal directly applied. At several establishments where gas is employed for certain industrial heat requirements attempts have been made to use the gas under boilers; at first glowing reports were circulated indicating a saving over coal of 20 %

and even $33\frac{1}{3}$ %. A little experience has always shown, however, not only that such results are not attained, but that the cost of the gaseous fuel is so much in excess of coal used directly as to make it necessary to return to the latter system.

" The following simple calculation will serve to show the uselessness of all attempts to convert bituminous coal into gas and distribute it to boiler plants.

"The average quality of fuel gas made from a trial run of several car loads of Illinois coal in a well-designed fuel gas plant showed a calorific value of 243,391 heat units per 1000 cubic feet of gas or 10,105,594 heat units per ton of coal. This is equivalent to 5052.8 heat units per pound of coal, whereas, by direct calorimeter test an average sample of the coal gave 11,172.6 heat units, or an efficiency of 45.2%.

" One pound of the coal by direct application showed a theoretical evaporation of 11.56 lbs. of water.

" The gas from one pound of the coal showed a theoretical evaporation of 5.23 lbs. water.

" 48.17 pounds of the coal were required to furnish 1000 feet of the gas.

" Taking the efficiency in the use of the coal direct at 50%,

" Taking the efficiency in the use of the gas direct at 90%,

" Taking the cost of the coal at 6 cents per bushel,

" Taking the cost of the gas at 8 cents per 1000 cubic feet,

we have as the cost of evaporating 1000 lbs. of water by coal directly applied :

" $\frac{50}{100}$ of 11.56 $=$ 5.78 lbs. of water to be evaporated in practice by 1 lb. of coal.

" $\frac{1000}{5.78}$ $=$ 173 lbs. of coal to evaporate 1000 lbs. of water ; 173 lbs. of coal at 6 cents per bushel $=$ 13 cents, and as the cost of the same coal converted into gas and applied,

" $\frac{90}{100}$ of 5.23 $=$ 4.71 lbs. of water to be evaporated in practice by gas from 1 lb. of coal.

" $\frac{1000}{4.71}$ $=$ 212.3 lbs. of coal required $=$ 4400 cubic feet of gas ; 4400 cubic feet of gas at 8 cents per 1000 $=$ 35.2 cents.

" It will be observed that the conditions assumed are especially favorable to the gas, the cost being placed at the remarkably low figure of 8 cents per 1000 cubic feet, which is about the actual cost of manufacture and distribution upon a large scale, and a very high efficiency is taken. Notwithstanding all this the coal used directly shows an advantage of over 170%."

NOTE: Prof. Potter's remarks are based on conditions found in St. Louis and adjacent territory.

Natural Gas, is variously constituted, and hence the estimates of its heating power vary.

Experiments in Pittsburgh show 1000 cubic feet of natural gas in actual efficiency under boilers equal to from 80 to 133 pounds coal. The coal varies from 12000 to 13000 B. T. U. per pound ; hence say 1,000,000 to 1,200,000 B. T. U. per 1000 feet of natural gas.

A Committee of the Western Society of Engineers of Pittsburgh, report 1 lb. good coal $=$ $7\frac{1}{2}$ cubic feet natural gas.

When burnt with just enough air its temperature of combustion is 4200° F. The Westinghouse Brake Co. in Pittsburgh found that with the best grade of Youghiogheny coal they could evaporate 10.38 lbs. water, and with the same boiler 1.18 feet natural gas evaporated 1 lb. water. They conclude that 1 lb. Youghiogheny coal $= 12\frac{1}{4}$ lb. natural gas, or 1000 cubic feet natural gas $= 81.6$ lb. coal.

The Indiana natural gas gives 1,100,000 B. T. U. for 1000 cubic feet and weighs 0.045 lbs. per cubic foot.

The analyses compare as follows:

TABLE NO. 32.

Analyses of Natural Gas.

	Pittsburgh, Pa., Gas.	Findlay, Ohio, Gas.
Hydrogen	22.0	2.18
Marsh gas	67.0	92.61
Carbonic oxide	0.6	0.26
Olefiant gas	1.0	0.30
Carbonic acid	0.6	0.50
Nitrogen	3.0	3.61
Oxygen	0.8	0.34
Ethylic hydride	5.0	----
Sulphuretted hydrogen	----	0.20
	100.0	100.00

500 H. P. Heine Boiler ready for transportation.

WATER.

Pure water at 62° F. weighs 62.355 pounds per cubic foot, or 8¼ pounds per U. S. gallon; 7.48 gallons = 1 cubic foot. It takes 30 pounds or 3.6 gallons for each horse-power per hour. It would be difficult to get at the total daily horse-power of steam used in the U. S., but it reaches into the billions of gallons of feed water per day.

The importance of knowing what impurities exist in most feed waters, how these act on a boiler, and how they may be removed is, therefore, patent to every intelligent engineer.

We give, therefore, the thoughts of some prominent investigators on the subject.

Prof. Thurston says:

"Incrustation and sediment are deposited in boilers, the one by the precipitation of mineral or other salts previously held in solution in the feed-water, the other by the deposition of mineral insoluble matters, usually earths, carried into it in suspension or mechanical admixture. Occasionally also vegetable matter of a glutinous nature is held in solution in the feed-water, and, precipitated by heat or concentration, covers the heating-surfaces with a coating almost impermeable to heat and hence liable to cause an over-heating that may be very dangerous to the structure. A powdery mineral deposit sometimes met with is equally dangerous, and for the same reason. *The animal and vegetable oils and greases carried over from the condenser or feed water heater are also very likely to cause trouble.* Only mineral oils should be permitted to be thus introduced, and that in minimum quantity. Both the efficiency and the safety of the boiler are endangered by any of these deposits.

" The amount of the foreign matter brought into the steam-boiler is often enormously great. A boiler of 100 horse-power uses, as an average, probably a ton and a half of water per hour, or not far from 400 tons per month, steaming ten hours per day, and even with water as pure as the Croton at New York, receives 90 pounds of mineral matter, and from many spring waters a *ton* which must be either blown out or deposited. These impurities are usually either calcium carbonate or calcium sulphate, or a mixture ; the first is most common on land, the second at sea. Organic matters often harden these mineral scales, and make them more difficult of removal.

" The only positive and certain remedy for incrustation and sediment once deposited is periodical removal by mechanical means, at sufficiently frequent intervals to insure against injury by too great accumulation. Between times, some good may be done by special expedients suited to the individual case. No one process and no one antidote will suffice for all cases.

City and County Building,
SALT LAKE CITY, UTAH.
Equipped with 225 H. P. Heine Safety Boilers.

"Where carbonate of lime exists, sal-ammoniac may be used as a preventive of incrustation, a double decomposition occuring, resulting in the production of ammonium carbonate and calcium chloride—both of which are soluble, and the first of which is volatile. The bicarbonate may be in part precipitated before use by heating to the boiling-point, and thus breaking up the salt and precipitating the insoluble carbonate. Solutions of caustic lime and metallic zinc act in the same manner. Waters containing tannic acid and the acid juices of oak, sumach, logwood, hemlock, and other woods, are sometimes employed, but are apt to injure the iron of the boiler, as may acetic or other acid contained in the various saccharine matters often introduced into the boiler to prevent scale, and which also make the lime-sulphate scale more troublesome than when clean. Organic matters should never be used.

"The sulphate scale is sometimes attacked by the carbonate of soda, the products being a soluble sodium sulphate and a pulverulent insoluble calcium carbonate, which settles to the bottom like other sediments and is easily washed off the heating-surfaces. Barium chloride acts similarly, producing barium sulphate and calcium chloride. All the alkalies are used at times to reduce incrustations of calcium sulphate, as is pure crude petroleum, the tannate of soda, and other chemicals.

"The effect of incrustation, and of deposits of various kinds, is to enormously reduce the conducting power of heating-surfaces ; so much so, that the power, as well as the economic efficiency of a boiler, may become very greatly reduced below that for which it is rated, and the supply of steam furnished by it may become wholly inadequate to the requirements of the case.

"It is estimated that a sixteenth of an inch (0.16 cm.) thickness of hard 'scale' on the heating-surface of a boiler will cause a waste of nearly one-eighth its efficiency, and the waste increases as the square of its thickness. The boilers of steam vessels are peculiarly liable to injury from this cause where using salt water, and the introduction of the surface-condenser has been thus brought about as a remedy. Land boilers are subject to incrustation by the carbonate and other salts of lime, and by the deposit of sand or mud mechanically suspended in the feed-water.

"It has been estimated that the annual cost of operation of locomotives in limestone districts is increased $750 by deposits of scale."

We give below an extract from an interesting paper on the " Impurities of Water," contributed by Messrs. Hunt and Clapp, to the transactions of the American Institute of Mining Engineers, for 1888.

Commercial Analyses.

By far the most common commercial analysis of water is made to determine its fitness for making steam. Water containing more than five parts per hundred thousand of free sulphuric or nitric acid is liable to cause serious corrosion, not only of the metal of the boiler itself, but of the pipes, cylinders, pistons, and valves with which the steam comes in contact. Sulphuric acid is the only one of these acids liable to be present in the water from natural sources; it being often produced in the water of the coal and iron districts, by the oxidation of iron pyrites to sulphate of iron, which, being soluble, is lixiviated from the earth strata, and carried into the stream. The presence of organic matter taken up by the water in its after-course, reducing the iron and lining the bottom of the stream with red oxide of iron, and leaving a considerable proportion of the sulphuric acid free in the water. This is a troublesome feature with the water necessarily used in many of the iron districts of this country. The sulphuric acid may come from other natural chemical reactions than the one described above. Muriatic and nitric acids, as well as often sulphuric acid, may be conveyed into water through the refuse of various kinds of manufacturing establishments discharged into it.

The large total residue in water used for making steam causes the interior linings of the boilers to become coated, clogs their action, and often produces a dangerous hard scale, which prevents the cooling action of the water from protecting the metal against burning.

Lime and magnesia bicarbonates in water lose their excess of carbonic acid on boiling, and often, especially when the water contains sulphuric acid, produce, with the other solid residues constantly being formed by the evaporation, a very hard and insoluble scale.

A larger amount than 100 parts per 100,000 of total solid residue will ordinarily cause troublesome scale, and should condemn the water for use in steam boilers, unless a better supply cannot be obtained.

The following is a tabulated form of the causes of trouble with water for steam purposes, and the proposed remedies, given by Prof. L. M. Norton, in his lecture on "Industrial Chemistry."

Brief Statement of Causes of Incrustation.

1. Deposition of suspended matter.
2. Deposition of dissolved salts from concentration.
3. Deposition of carbonates of lime and magnesia by boiling off carbonic acid, which holds them in solution.
4. Deposition of sulphates of lime, because sulphate of lime is but slightly soluble in cold water, less soluble in hot water, insoluble above 140° Centigrade. (284 degrees Fahrenheit.)
5. Deposition of magnesia, because magnesium salts decompose at high temperature.
6. Deposition of lime soap, iron soap, etc., formed by saponification of grease.

Various Means of Preventing Incrustation.

1. Filtration.
2. Blowing off.
3. Use of internal collecting apparatus or devices for directing the circulation.
4. Heating feed water.
5. Chemical or other treatment of water in boiler.
6. Introduction of zinc into boiler.
7. Chemical treatment of water outside of boiler.

Tabular View.

Troublesome Substance.	Trouble.	Remedy or Palliation.
Sediment, mud, clay, etc.	Incrustation.	Filtration. Blowing off.
Readily soluble salts.	Incrustation.	Blowing off.
Bicarbonates of lime, magnesia, iron.	Incrustation.	Heating feed. Addition of caustic soda, lime, or magnesia, etc.
Sulphate of lime.	Incrustation.	Addition of carbonate of soda, barium chloride.
Chloride and sulphate of magnesium.	Corrosion.	Addition of carbonate of soda, etc.
Carbonate of soda in large amounts.	Priming.	Addition of barium chloride, etc.
Acid (in mine waters).	Corrosion.	Alkali.
Dissolved carbonic acid and oxygen.	Corrosion.	Heating feed. Addition of caustic soda, slacked lime, etc.
Grease (from condensed water).	Corrosion.	Slacked lime and filtering. Carbonate of soda. Substitute mineral oil.
Organic matter (sewage).	Priming.	Precipitate with alum or ferric chloride and filter.
Organic matter.	Corrosion.	Ditto.

The mineral matters causing the most troublesome boiler-scales are bi carbonates and sulphates of lime and magnesia, oxides of iron and alumina, and silica. We present here a table showing the amount and nature of impurities in feed water in different sections of the United States. (Table 33.)

NOTE. The mud drum of the Heine Boiler, surrounded as it is, by water at a temperature of about 350° F., forms a sort of live steam purifier in which a large part of the scale forming salts are precipitated. It is largely on this account that the Heine Boiler is able to work satisfactorily with the most impure waters, where other boilers, lacking the mud-drum-purifier, fail of success altogether. This has been practically demonstrated on many occasions. Probably no "tougher" water is encountered by boiler users anywhere, than in Columbus, Ohio. Heine Boilers supplanted flue boilers there, that were struggling in vain against scale. The success of the Heine Boiler with this water was a most unqualified one. The L. Hoster Brewing Co. and the Columbus Electric Light and Power Co. both have large plants of Heine Boilers, and we think will cheerfully testify to the superiority of the Heine Boiler in this respect. It is not claimed that NO scale will form in the Heine Boiler when operated with scale producing water. It is only those boilers which have no particular reputation for good service, those boilers that are guaranteed (?) to do anything and everything, that run scaleless on bad water. Eternal vigilance is the price of many things besides liberty and constant watchfulness is necessary if scale is to be avoided in any boiler. But common, every day experience has shown that the conditions which aid in the prevention of scale in boilers are more perfectly provided for in the Heine than in any other type.

Oil or grease often causes as much trouble in boilers as scale or mud, and is much more difficult to remove, as it cannot be "blown off." It requires especial care where a part or the whole of the feed water comes from condensers or from heating coils where exhaust steam is used.

We reprint a warning given by the oldest boiler insurance company in the United States.

Table of Water Analyses.

Grains per U. S. Gallon, 231 Cubic Inches.

WHERE FROM.	Lime and Magnesia Carbonates.	Lime and Magnesia Sulphates.	Sodium Chloride. (Salt.)	Iron Oxide, Carb. Sulph., etc.	Volatile and Organic Matter.	Total Solids in Grains.
Buffalo, N. Y., Lake Erie	5.66	3.32	0.58	------	0.18	9.74
Pittsburgh, Allegheny River	0.37	3.78	0.58	0.37	1.50	6.00
Pittsburgh, Monongahela River	1.06	5.12	0.64	0.78	3.20	10.80
Milwaukee, Wisconsin River	6.23	4.67	1.76	20.14	6.50	39.30
Galveston, Texas, 1	13.68	13.52	326.64	Trace.	Trace.	353.84
Columbus, Ohio	20.76	11.74	7.02	0.58	6.50	46.60
Washington, D. C., city supply	2.87	3.27	Trace.	0.36	2.10	8.60
Baltimore, Md., city supply	2.77	0.65	Trace.	0.10	3.80	7.30
Sioux City, Ia., city supply	19.76	1.24	1.17	1.03	4.40	27.60
Los Angeles, Cal., 1	10.12	5.84	3.51	2.63	4.10	26.20
Bay City, Michigan, Bay	8.47	10.36	20.48	1.15	8.74	49.20
Bay City, Michigan, River	4.84	33.66	126.78	3.00	10.92	179.20
Cincinnati, Ohio River	3.88	0.78	1.79	------	Trace.	6.73
Watertown, Conn.	1.47	4.51	1.76	Trace.	1.78	9.52
Ft. Wayne, Ind.	8.78	6.22	3.51	1.59	10.98	31.08
Wilmington, Del.	10.04	6.02	4.29	8.48	6.17	35.00
Galveston, Texas, 2	21.79	29.149	398.99	------	4.00	453.93
Wichita, Kansas	14.14	25.91	24.34	------	2.00	66.39
Los Angeles, Cal., 2	3.72	12.59	------	0.76	6.00	23.07
St. Louis, Mo., well water	27.04	23.73	15.57	3.49	0.46	70.29
Pittsburgh, Pa., artesian well	23.45	5.71	18.41	1.04	0.82	49.43
Springfield, Ill., 1	12.90	7.40	1.97	2.19	8.62	33.17
Springfield, Ill., 2	5.47	4.31	1.56	4.28	5.83	21.45
Hillsboro, Ill.	14.56	2.97	2.39	1.63	Trace.	21.55
Pueblo, Colo.	4.32	16.15	1.20	1.97	5.12	28.76
Long Island City, L. I.	4.0	28.0	16.0	------	1.0	39.0
Mississippi River, above Missouri River	8.24	1.02	0.50	------	5.25	15.01
Mississippi River, below mouth of Missouri River	10.64	7.41	1.36	1.22	15.86	36.49
Mississippi River at St. Louis W. W.	9.64	6.94	1.54	1.57	9.85	29.54
Missouri River above mouth	10.07	8.92	1.87	3.26	11.37	35.49

(Reprinted from "THE LOCOMOTIVE," March, 1885; published by the Hartford Steam Boiler Inspection and Insurance Co.)

The Effect of Oil in Boilers.

We have often referred to the fact that the presence of grease or any of the animal oils in steam boilers is almost certain to cause trouble. Our illustration this month gives a better idea of the effect produced than pages of verbal description possibly could. It is from a photograph and is nowise exaggerated

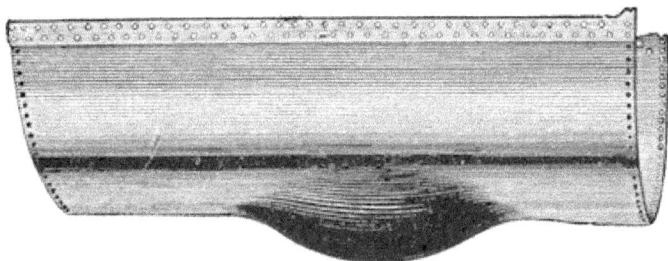

The boiler from which the plate shown in the cut was taken, was a nearly new one. It was made of a well-known brand of mild steel, and that it was admirably adapted to the purpose for which it was used, is proved by its stretching as it did without rupture. The dimensions of bulge shown are four feet lengthwise of the boiler, three feet girthwise and nine inches deep. The metal, originally 5-16 of an inch thick, drew down to $\frac{1}{8}$ inch in thickness at the lowest point of the "bag" without the slightest indication of fracture.

The circumstances under which the bulge occurred may best be described in the words of the inspector who examined the boiler, and are as follows:

"Last Tuesday morning I was called in great haste to the————works. Upon arrival I found one of the boilers badly bulged, and with twenty pounds of steam up. I could give no explanation until I had thoroughly examined the internal parts of the boiler. I gave directions for cooling the boiler and ordered top man-hole plate to be loosened, but not to be taken out until my arrival in the afternoon, that I might see everything undisturbed. This was done. On my arrival I took out the man-hole plates in top of shell and front head * * * and made an examination."

"I found that the boiler had been cleaned the preceding Sunday, and at that time a gallon or more of black oil had been thrown into it. Monday morning the boiler was fired up and was run through the day at a pressure of 90 pounds per square inch. At six o'clock Monday night the engine was stopped, the drafts were closed, and no more firing was done until nine o'clock. Upon going to fire up at this time, the bulge was observed. From six to nine o'clock a pressure of only 40 pounds was carried."

"Upon examination I found the entire boiler saturated with this oil."

This is almost certain to be the result of putting grease into a steam boiler. It settles down on the fire-sheets, when the draft is closed, and the circulation of water nearly stops, and prevents contact between the plates and the water. As a consequence, the plates over the fire become over-heated; and under such circumstances a very slight steam-pressure is sufficient to bag the sheets. Unless the boiler is made of very good material, the plate is apt to be fractured, and explosion is likely to occur.

When oil is used to remove scale from steam-boilers, too much care cannot be exercised to make sure that it is free from grease or animal oil. Nothing but pure mineral oil should be used. Crude petroleum is one thing; black oil, which may mean almost anything, is very likely to be something quite different.

The action of grease in a boiler is peculiar, but not more so than we might expect. It does not dissolve in the water, nor does it decompose, neither does it remain on top of the water, but it seems to form itself into what may be described as "slugs," which at first seem to be slightly lighter than the water, of just such a gravity, in fact, that the circulation of the water carries them about at will. After a short season of boiling, these "slugs" or suspended drops seem to acquire a certain degree of "stickiness," so that when they come in contact with shell and flues of the boiler, they begin to adhere thereto. Then under the action of heat they begin the process of "varnishing" the interior of the boiler. *The thinnest possible coating of this varnish is sufficient to bring about overheating of the plates*, as we have found repeatedly in our experience. We emphasize the point that it is *not* necessary to have a coating of grease of any appreciable thickness to cause overheating and bagging of plates and leakage at seams.

The time when damage is most likely to occur is after the fires are banked, for then, the formation of steam being checked, the circulation of water stops, and the grease thus has an opportunity to settle on the bottom of the boiler and prevent contact of the water with the fire-sheets. Under these circumstances, a very low degree of heat in the furnace is sufficient to overheat the plates to such an extent that bulging is sure to occur. When the facts are understood, it will be found quite unnecessary to attribute the damage to low water.

This accident also serves to illustrate the perfection to which the manufacture of steel for boiler plates has attained. It would be an extraordinarily good quality of iron that would stand such a test without fracture.

250 H. P. Heine Boiler "en route."

New Planters House,
ST. LOUIS, MO.
Contains 800 H. P. of Heine Boilers.

Weight of Water.

The weight of water varies with the temperature as given by the following table. (C. A. SMITH.)

TABLE NO. 34.

Weight of One Cubic Foot Water at Various Temperatures.

Temp., Degrees F.	Weight per Cubic Foot.	Temp., Degrees F.	Weight per Cubic Foot.	Temp., Degrees F.	Weight per Cubic Foot.	Temp., Degrees F.	Weight per Cubic Foot.
32	62.418	85	62.182	145	61.291	205	59.930
35	62.422	90	62.133	150	61.201	210	59.820
39.1	62.425	95	62.074	155	61.096	212 By formula.	59.760
40	62.425	100	62.022	160	60.991	212 By measurem't	59.640
45	62.422	105	61.960	165	60.843	230	59.360
50	62.409	110	61.868	170	60.783	250	58.780
55	62.394	115	61.807	175	60.665	270	58.150
60	62.372	120	61.715	180	60.548	290	57.590
65	62.344	125	61.654	185	60.430	298	57.270
70	62.313	130	61.563	190	60.314	338	56.140
75	62.275	135	61.472	195	60.198	366	55.290
80	62.232	140	61.381	200	60.081	390	54.540

Very often in the trials of a boiler or engine the most convenient unit of measurement of water is the cubic foot. This will be the case when a weir measurement is made or when the water is measured by a water meter. The use of a water meter involves many precautions, the most important being the following: The meter should work under moderate head of supply and small head of delivery; it should be set in such a manner that it can be tested in place under the exact conditions of use; if a positive meter, it should be especially constructed to work freely, if it is to be used in warm water. This table is also used for estimating the weight of water in boilers, and for correcting boiler trials for differences of water level.

150 H. P. Heine Boiler.

Size for Water Pipes.

We found at beginning of this article, 3.6 gallons feed water are required for each H. P. per hour. This makes 6 gallons per minute for a 100 H. P. boiler. In proportioning pipes, however, it is well to remember that boiler work is seldom perfectly steady, and that as the engine cuts off just as much steam as the work demands at each stroke, *all the discrepancies of demand and supply have to be equalized in the boiler.* Therefore we may often have to evaporate during one-half hour 50 to 75 per cent more than the normal requirements. For this reason it is sound policy to arrange the feed pipes so that 10 gallons per minute may flow through them, without undue speed or friction, for each 100 H. P. of boiler capacity. The following tables will facilitate this work:

TABLE NO. 35.

Table Giving Rate of Flow of Water, in Ft. per Min., Through Pipes of Various Sizes, for Varying Quantities of Flow.

Gallons per min.	¾″	1″	1¼″	1½″	2″	2½″	3″	4″
5	218	122½	78½	54½	30½	19½	13½	7⅗
10	436	245	157	109	61	38	27	15⅕
15	653	367½	235½	163½	91½	58½	40½	23
20	872	490	314	218	122	78	54	30⅗
25	1090	612½	392½	272½	152½	97½	67½	38⅕
30	-------	735	451	327	183	117	81	46
35	-------	857½	549½	381½	213½	136½	94½	53⅗
40	-------	980	628	436	244	156	108	61⅕
45	-------	1102½	706½	490½	274½	175½	121½	69
50	-------	-------	785	545	305	195	135	76⅗
75	-------	-------	1177½	817½	457½	292½	202½	115
100	-------	-------	-------	1090	610	380	270	153⅕
125	-------	-------	-------	-------	762½	487½	337½	191⅗
150	-------	-------	-------	-------	915	585	405	230
175	-------	-------	-------	-------	1067½	682½	472½	268⅕
200	-------	-------	-------	-------	1220	780	540	306⅗

TABLE NO. 36.

Table Giving Loss in Pressure Due to Friction, in Pounds per Sq. In., for Pipe 100 Ft. Long.

By G. A. Ellis, C. E.

Gallons discharged per min.	¾″	1″	1¼″	1½″	2″	2½″	3″	4″
5	3.3	0.84	0.31	0.12	-------	-------	-------	-------
10	13.0	3.16	1.05	0.47	0.12	-------	-------	-------
15	28.7	6.98	2.38	0.97	-------	-------	-------	-------
20	50.4	12.3	4.07	1.66	0.42	-------	-------	-------
25	78.0	19.0	6.40	2.62	-------	0.21	0.10	-------
30	-------	27.5	9.15	3.75	0.91	-------	-------	-------
35	-------	37.0	12.4	5.05	-------	-------	-------	-------
40	-------	48.0	16.1	6.52	1.60	-------	-------	-------
45	-------	-------	20.2	8.15	-------	-------	-------	-------
50	-------	-------	24.9	10.0	2.44	0.81	0.35	0.09
75	-------	-------	56.1	22.4	5.32	1.80	0.74	-------
100	-------	-------	-------	39.0	9.46	3.20	1.31	0.33
125	-------	-------	-------	-------	14.9	4.89	1.99	-------
150	-------	-------	-------	-------	21.2	7.0	2.85	0.69
175	-------	-------	-------	-------	28.1	9.46	3.85	-------
200	-------	-------	-------	-------	37.5	12.47	5.02	1.22

Loss of Head Due to Bends.

Bends produce a loss of head in the flow of water in pipes. Weisbach gives the following formula for this loss :

$H = f \frac{v^2}{2g}$ where H = loss of head in feet, f = coefficient of friction, v = velocity of flow in feet per second, g = 32.2.

As the loss of head or pressure is in most cases more conveniently stated in pounds per square inch, we may change this formula by multiplying by 0.433, which is the equivalent in pounds per square inch for one foot head.

If P = loss in pressure in pounds per square inch, F = coefficient of friction.

$P = F \frac{v^2}{64.4}$, v being the same as before.

From this formula has been calculated the following table of values for F, corresponding to various exterior angles, A.

TABLE NO. 37.

A =	20°	40°	45°	60°	80°	90°	100°	110°	120°	130°
F =	0.020	0.060	0.079	0.158	0.320	0.426	0.546	0.674	0.806	0.934

This applies to such short bends as are found in ordinary fittings, such as 90° and 45° Ells, Tees, etc.

A globe valve will produce a loss about equal to two 90° bends, a straightway valve about equal to one 45° bend. To use the above formula *find the speed p. second, being one-sixtieth of that found in Table No. 35; square this speed, and divide the result by 64.4; multiply the quotient by the tabular value of F corresponding to the angle of the turn, A.*

For instance a 400 H. P. battery of boilers is to be fed through a 2" pipe. Allowing for fluctuations we figure 40 gallons per minute, making 244 feet per minute speed, equal to a velocity of 4.06 feet per second. Suppose our pipe is in all 75 feet long; we have from Table No. 36, for 40 gallons per minute, 1.60 pounds loss; for 75 feet we have only 75 per cent. of this = 1.20 pounds. Suppose we have 6 right angled ells, each giving F = 0.426. We have then 4.06×4.06 = 16.48; divide this by 64.4 = 0.256. Multiply this by F = 0.426 pounds, and as there are six ells, multiply again by 6, and we have 6×0.426×0.256 = 0.654. The total friction in the pipe is therefore 1.20+0.654 = 1.854 pounds per square inch. If the boiler pressure is 100 pounds and the water level in the boiler is 8 feet higher than the pump suction level, we have first 8×0.433 = 3.464 pounds. The total pressure on the pump plunger then is 100+3.464+1.854 = 105.32 pounds per square inch. If in place of six right angled ells we had used three 45° ells, they would have cost us only 3×0.079 = 0.237 pounds; 0.237×0.256 = 0.061.

The total friction head would have been 1.20+0.061=1.261 and the total pressure on the plunger 100+3.464+1.261=104.73 pounds per square inch, a saving over the other plan of nearly 0.6 pounds.

To be accurate, we ought to add a certain head in either case " to produce the velocity." But this is very small, being for velocities of :

2; 3; 4; 5; 6; 8; 10; 12 and 18 feet per sec. 0.027 ; 0.061; 0.108; 0.168; 0.244; 0.433; 0.672; 0.970 and 2.18 lbs. per sq. in. Our results should therefore have been increased by about 0.11 lbs.

Foresters' Temple.
Headquarters of Independent Order of Foresters,
TORONTO, ONT., CANADA.
Contains 240 H. P. of Heine Boilers.

It is usual, however, to use larger pipes and thus to materially reduce the frictional losses.

Rating Boilers by Feed Water.

The rating of boilers has, since the Centennial in 1876, been generally based on 30 pounds feed water per hour per H. P. This is a fair average for good non-condensing engines working under about 70 to 100 pounds pressure. But different pressures and different rates of expansion change the requirements for feed-water. The following table, No. 38, gives Prof. R. H. Thurston's estimate of the steam consumption for the *best classes of engines in common use, when of moderate size and in good order:*

TABLE NO. 38.

Weights of Feed Water and of Steam.

Non-condensing Engines.—R. H. T.

STEAM PRESSURE.		LBS. PER H. P. PER HOUR.—RATIO OF EXPANSION.					
Atmospheres.	Lbs. per sq. in.	2	3	4	5	7	10
3	45	40	39	40	40	42	45
4	60	35	34	36	36	38	40
5	75	30	28	27	26	30	32
6	90	28	27	26	25	27	29
7	105	26	25	24	23	25	27
8	120	25	24	23	22	22	21
10	150	24	23	22	21	20	20

Condensing Engines.

2	30	30	28	28	30	35	40
3	45	28	27	27	26	28	32
4	60	27	26	25	24	25	27
5	75	26	25	25	23	22	24
6	90	26	24	24	22	21	20
8	120	25	23	23	22	21	20
10	150	25	23	22	21	20	19

Small engines having greater proportional losses in friction, in leaks, in radiation, etc., and besides receiving generally less care in construction and running than larger ones, require more feed-water (or steam) per hour.

Table No. 39 gives Mr. R. H. Buel's estimate for such engines.

Feed-Water Required by Small Engines.

Pressure of Steam in Boiler, by Gauge.	Pounds of Water per Effective Horse-power per Hour.	Pressure of Steam in Boiler, by Gauge.	Pounds of Water per Effective Horse-power per Hour.
10	118	60	75
15	111	70	71
20	105	80	68
25	100	90	65
30	95	100	63
40	84	120	61
50	79	150	58

Boiler Room Alleghany Traction Co. Plant,
PITTSBURGH, PA.
500 H. P. Heine Boilers.

Heating Feed-Water.

Feed-water as it comes from wells or hydrants has ordinarily a temperature of from 35° in winter to from 60° to 70° in summer.

Much fuel can be saved by heating this water by the exhaust steam, whose heat would otherwise be wasted. Until quite recently, only non-condensing engines utilized feed-water heaters; but lately they have been introduced with success between the cylinder and the air pump in condensing engines. The saving in fuel due to heating feed-water is given in Table No. 40.

Percentage of Saving in Fuel by Heating Feed-Water. Steam at 70 Pounds Gauge Pressure.

Initial Temperature Feed.	TEMPERATURE TO WHICH FEED IS HEATED.														
	100°	110°	120°	130°	140°	150°	160°	170°	180°	190°	200°	210°	220°	250°	300°
35°	5.53	6.38	7.24	8.09	8.95	9.89	10.66	11.52	12.38	13.24	14.09	14.95	15.81	19.40	29.34
40°	5.12	5.97	6.84	7.69	8.56	9.42	10.28	11.14	12.00	12.87	13.73	14.59	15.45	18.89	28.78
45°	4.71	5.57	6.44	7.30	8.16	9.03	9.90	10.76	11.62	12.49	13.36	14.22	15.09	18.37	28.22
50°	4.30	5.16	6.03	6.89	7.76	8.64	9.51	10.38	11.24	12.11	12.98	13.85	14.72	17.87	27.67
55°	3.89	4.75	5.63	6.49	7.37	8.24	9.11	9.99	10.85	11.73	12.60	13.48	14.35	18.38	27.12
60°	3.47	4.34	5.21	6.08	6.96	7.84	8.72	9.60	10.47	11.34	12.22	13.10	13.98	16.86	26.56
65°	3.05	3.92	4.80	5.67	6.56	7.44	8.32	9.20	10.08	10.96	11.84	12.72	13.60	16.35	26.02
70°	2.62	3.50	4.38	5.26	6.15	7.03	7.92	8.80	9.68	10.57	11.45	12.34	13.22	15.84	25.47
75°	2.19	3.07	3.96	4.84	5.73	6.62	7.51	8.40	9.28	10.17	11.06	11.95	12.84	15.33	24.92
80°	1.76	2.65	3.54	4.42	5.32	6.21	7.11	8.00	8.88	9.78	10.67	11.57	12.46	14.82	24.37
85°	1.30	2.22	3.11	4.00	4.90	5.80	6.70	7.59	8.48	9.38	10.28	11.18	12.07	14.32	23.82
90°	0.89	1.78	2.68	3.58	4.48	5.38	6.28	7.18	8.07	8.98	9.88	10.78	11.68	13.81	23.27
95°	0.45	1.34	2.25	3.15	4.05	4.96	5.86	6.77	7.66	8.57	9.47	10.38	11.29	13.31	22.73
100°	0.00	0.90	1.81	2.71	3.62	4.53	5.44	6.35	7.25	8.16	9.07	9.98	10.88	12.80	22.18

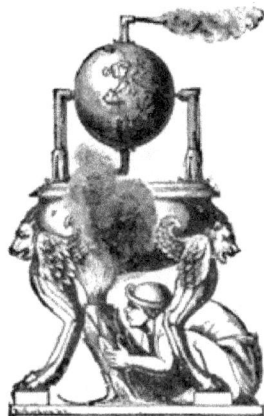

STEAM.

When water is heated in an open vessel its temperature rises until it reaches 212° (at sea level); if more heat is added a portion of the water changes from a liquid form to a vapor called *steam*. If the process is carried on in a closed vessel the pressure within the same rises on account of the expansive force of the steam. The water then will rise to a higher temperature with each increment of pressure before it begins to boil and form steam.

For the distinction between "sensible" and "latent" heat see p. 7.

The following table No. 41, giving the properties of saturated steam, is adapted from Prof. Peabody's well known tables. The first column gives the actual pressure in pounds per square inch above the atmosphere.

Column two gives the temperature in degrees Fahrenheit for the corresponding pressure.

Columns three and four give the heat, in heat units, of steam and water, respectively, from 32° F.

Column five gives the heat of vaporization for the corresponding pressure, and is the difference between columns three and four.

Columns six and seven give the weight of one cubic foot in pounds and the volume of one pound in cubic feet, of saturated steam.

Column eight gives the approximate weight of one cubic foot of water for the corresponding weight and temperature and is calculated from Prof. Rankin's approximate formula :

$$D = \frac{2 D_0}{\dfrac{T_0 + 461}{500} + \dfrac{500}{T_0 + 461}} \quad \text{where}$$

D = required density. D_0 = max. density = 62.425 lbs.
T_0 = given temperature in degrees F.

Column nine gives the factor of equivalent evaporation from and at 212° F., assuming feed to be 212° in each case. For the factor of evaporation for any temperature of feed, add 0.00104 to the given factor for each degree difference in temperature between feed and 212°.

For complete table of factors of evaporation, see page 152.

The horse-power of a boiler is obtained by dividing the equivalent evaporation from and at 212° by 30.978. This is on the basis of feed from 212° to steam at 70 pounds pressure. On the basis of feed from 100° to steam at 70 lbs., divide the equivalent evaporation by 34.485.

TABLE NO. 41.

Table of the Properties of Saturated Steam.

From Peabody's Tables.

1	2	3	4	5	6	7	8	9
Gauge Pressure in lbs. per Square Inch.	Temperature in Degrees F.	Total Heat in Heat Units from Water at 32° F.	Heat Units in Liquid from 32° F.	Heat of Vaporization in Heat Units.	Density or Weight of 1 Cu. ft. in lbs.	Volume of 1 lb. in Cubic Feet.	Weight of 1 Cubic ft. of Water.	Factor of Equivalent Evaporation from and at 212° F.
0	212.00	1146.6	180.8	965.8	0.03760	26.60	59.76 (Formula) 59.64 (Observed)	1.0000
10	239.36	1154.9	208.4	946.5	0.06128	16.32	59.04	1.0086
20	258.68	1160.8	227.9	932.9	0.08439	11.85	58.50	1.0147
30	273.87	1165.5	243.2	922.3	0.1070	9.347	58.07	1.0196
40	286.54	1169.3	255.9	913.4	0.1292	7.736	57.69	1.0235
50	297.46	1172.6	266.9	905.7	0.1512	6.612	57.32	1.0269
55	302.42	1174.2	271.9	902.3	0.1621	6.169	57.22	1.0286
60	307.10	1175.6	276.6	899.0	0.1729	5.784	57.08	1.0300
65	311.54	1176.9	281.1	895.8	0.1837	5.443	56.95	1.0314
70	315.77	1178.2	285.6	892.7	0.1945	5.142	56.82	1.0327
75	319.80	1179.5	289.8	889.8	0.2052	4.873	56.69	1.0341
80	323.66	1180.6	293.8	886.9	0.2159	4.633	56.59	1.0352
85	327.36	1181.8	297.7	884.2	0.2265	4.415	56.47	1.0365
90	330.92	1182.8	301.5	881.5	0.2371	4.218	56.36	1.0375
95	334.35	1183.9	305.0	879.0	0.2477	4.037	56.25	1.0386
100	337.66	1184.9	308.5	876.5	0.2583	3.872	56.18	1.0397
105	340.86	1185.9	311.8	874.1	0.2689	3.720	56.07	1.0407
110	343.95	1186.8	315.0	871.8	0.2794	3.580	55.97	1.0417
115	346.94	1187.7	318.2	869.6	0.2898	3.452	55.87	1.0426
120	349.85	1188.6	321.2	867.4	0.3003	3.330	55.77	1.0435
125	352.68	1189.5	324.2	865.3	0.3107	3.219	55.69	1.0444
130	355.43	1190.3	327.0	863.3	0.3212	3.113	55.58	1.0452
135	358.10	1191.1	329.8	861.3	0.3315	3.017	55.52	1.0461
140	360.70	1191.9	332.5	859.4	0.3420	2.924	55.44	1.0469
145	363.25	1192.8	335.2	857.5	0.3524	2.838	55.36	1.0478
150	365.73	1193.5	337.8	855.7	0.3629	2.756	55.29	1.0486
155	368.62	1194.3	340.3	853.9	0.3731	2.681	55.22	1.0494
160	370.51	1195.0	342.8	852.1	0.3835	2.608	55.15	1.0500
165	372.83	1195.7	345.2	850.4	0.3939	2.539	55.07	1.0508
170	375.09	1196.3	347.6	848.7	0.4043	2.474	54.99	1.0514
175	377.31	1197.0	349.9	847.1	0.4147	2.412	54.93	1.0522
180	379.48	1197.7	352.2	845.4	0.4251	2.353	54.86	1.0529
185	381.60	1198.3	354.4	843.9	0.4353	2.297	54.79	1.0535
190	383.70	1199.0	356.6	842.3	0.4455	2.244	54.73	1.0542
195	385.75	1199.6	358.8	840.8	0.4559	2.193	54.66	1.0549
200	387.76	1200.2	360.9	839.2	0.4663	2.145	54.60	1.0555
225	397.36	1203.1	370.9	832.2	0.5179	1.930	54.27	1.0585
250	406.07	1205.8	380.1	825.7	0.5699	1.755	54.03	1.0613
275	414.22	1208.3	388.5	819.8	0.621	1.609	53.77	1.0639
300	421.83	1210.6	396.5	814.1	0.674	1.483	53.54	1.0666

— 72 —

The Betz Building,
PHILADELPHIA, PA.,
Contains 500 H. P. Heine Boilers.

Of the Motion of Steam.

The flow of steam of a greater pressure into an atmosphere of a less pressure, increases as the difference of pressure is increased, until the external pressure becomes only 58 per cent of the absolute pressure in the boiler. The flow of steam is neither increased nor diminished by the fall of the external pressure below 58 per cent, or about $\frac{4}{7}$ths of the inside pressure, even to the extent of a perfect vacuum. In flowing through a nozzle of the best form, the steam expands to the external pressure, and to the volume due to this pressure, so long as it is not less than 58 per cent of the internal pressure. For an external pressure of 58 per cent, and for lower percentages, the ratio of expansion is 1 to 1.624. The following table, No. 42, is selected from Mr. Brownlee's data exemplifying the rates of discharge, under a constant internal pressure, into various external pressures:

TABLE NO. 42.

Outflow of Steam ; From a Given Initial Pressure into Various Lower Pressures.

Absolute Initial Pressure in Boiler, 75 Lbs. per Square Inch.

D. K. C.

Absolute Pressure in Boiler in Lbs. per Square Inch.	External Pressure in Lbs. per Square Inch.	Ratio of Expansion in Nozzle.	Velocity of Outflow at Constant Density.	Actual Velocity of Outflow. Expanded.	Discharge per Square Inch of Orifice per Minute.
Lbs.	Lbs.	Ratio.	Ft. per Sec.	Ft. per Sec.	Lbs.
75	74	1.012	227.5	230.	16.68
75	72	1.037	386.7	401.	28.35
75	70	1.063	490.	521.	35.93
75	65	1.136	660.	749.	48.38
75	61.62	1.198	736.	876.	53.97
75	60	1.219	765.	933.	56.12
75	50	1.434	873.	1252.	64.
75	45	1.575	890.	1401.	65.24
75	43.46 (58%)	1.624	890.6	1446.5	65.3
75	15	1.624	890.6	1446.5	65.3
75	0	1.624	890.6	1446.5	65.3

When, on the contrary, steam of varying initial pressure is discharged into the atmosphere—pressures of which the atmospheric pressure is not more than 58 per cent—the velocity of outflow at constant density, that is, supposing the initial density to be maintained, is given by the formula—

$$V = 3.5953 \sqrt{h} \quad (1)$$

where V = the velocity of outflow in feet per minute, as for steam of the initial density. h = the height in feet of a column of steam of the given absolute initial pressure of uniform density, the weight of which is equal to the pressure on the unit of base.

The following table is calculated from this formula :

Outflow of Steam into the Atmosphere.

D. K. C.

Absolute initial pressure in lbs. per sq. in.	External pressure in lbs. per sq. in.	Ratio of expansion in nozzle.	Velocity of outflow at constant density.	Actual velocity of outflow, expanded.	Discharge per sq. in. of orifice per min.
Lbs.	Lbs.	Ratio.	Ft. per sec.	Ft. per sec.	Lbs.
25.37	14.7	1.624	863	1401	22.81
30	14.7	1.624	867	1408	26.84
40	14.7	1.624	874	1419	35.18
45	14.7	1.624	877	1424	39.78
50	14.7	1.624	880	1429	44.06
60	14.7	1.624	885	1437	52.59
70	14.7	1.624	889	1444	61.07
75	14.7	1.624	891	1447	65.30
90	14.7	1.624	895	1454	77.94
100	14.7	1.624	898	1459	86.34
115	14.7	1.624	902	1466	98.76
135	14.7	1.624	906	1472	115.61
155	14.7	1.624	910	1478	132.21
165	14.7	1.624	912	1481	140.46
215	14.7	1.624	919	1493	181.58

The Economic Value of Dry Steam.

Saturated steam is defined as steam of the maximum pressure and density due to its temperature. It is steam in its normal condition, being both at the condensing and the generating point. It is formed thus in a well-designed boiler, and any heat added would evaporate more water, while heat taken away would condense some of the steam. In badly-proportioned boilers, however, we find water entrained in the steam in the form of a fine mist. This is caused by imperfect arrangements for separating the steam from the water; by a liberating surface either too small or too near the hot metal; by a cramped or low steam-space; or by more heating surface than the water-space or circulation warrants. It is only during the last decade that the attention of steam users generally has been bent on getting *dry steam, i.e.*, saturated steam containing but a small percentage of entrained water.

Formerly, with long stroke and slow speed engines, and when cylinder condensation was understood but by a few experts, this entrainment was rarely measured.

In Mr. D. K. Clark's celebrated Manual for Mechanical Engineers (1877), which contains the record and careful analysis of many notable boiler tests, entrainment is not even mentioned. Most of the high results of ancient tests which are paraded in advertisements are therefore open to the suspicion that they may have been obtained by delivering " soda water " in place of steam. Since calorimeter tests have become common, entrainments up to 6 and 10 per cent. have been found in boilers apparently giving high economy. As early as 1860, Chief Engineer Isherwood, of the U. S. Navy, began investigating the economic losses due to moisture in the cylinder.

Superheated steam was suggested as a remedy for cylinder condensation by Prof. Dixwell, of Boston, early in 1875, and Mr. Hirn, of Mulhouse, made extensive and successful experiments in this line in 1873 and 1875 (first published in 1877). Where good saturated steam induces such wasteful condensation in the cylinder, wet steam greatly increases the losses. For the water cools the internal surfaces of the cylinder more rapidly than steam of the same temperature, and this increases the cylinder condensation. Hence, economic reasons condemned wet steam, and finally close-coupled and high-speed engines protested against entrainment in the emphatic language of broken valves and blown out cylinder heads.

Marine boilers are called upon for a maximum of work in a minimum of space, and are therefore more liable to entrain water; this was especially the case with the low-pressures in use before 1880. We therefore find super-heated steam resorted to in the navy at an early day.

Exhaustive experiments made by Mr. Isherwood early in the sixties show large gains in economy by superheating, and thus illustrate the losses due to water in the steam.

We choose only two examples in which the boiler pressure and the rate of expansion are alike; the economy found is therefore clearly due to super-heating the steam, or conversely the loss is due to cylinder condensation.

TABLE NO. 44.

NAME OF STEAMER.	Pounds Gauge Pressure.	Rate of Expansion.	Pounds Coal Per H.P. per h.	Character of steam used.	Saving in Coal.
Dallas	32	3.22	3.80	Saturated.	----------
Georgeanna	33	3.22	2.58	Superheated.	47.3 %
Eutaw	27	1.85	3.84	Saturated.	----------
Eutaw	28	1.85	2.99	Superheated.	28.4 %

At the instance of Prof. Dixwell, the Government in 1877 sent Chief Engineers Loring, Baker and Farmer to Boston to test the effect of super-heated steam on the small Corliss engine of the Institute of Technology. The boiler pressure throughout the six tests was kept uniform. Three different rates of expansion were taken, and with each, one test was run with saturated and one with superheated steam, the degree of superheat being adjusted to the rate of expansion. The total steam used was condensed and weighed, and the loss by cylinder condensation thus accurately determined.

TABLE NO. 45.

Tests of Corliss Engine 8" × 24," Mass. Inst. of Technology.

Pounds Boiler Pressure.	Rate of Expansion.	Superheat.	Pounds steam per H. P. per hour.		Loss by moisture when using Saturated Steam.
			1st Test. Superheated.	2d Test. Saturated.	
50.4	4.05	279° F.	19.39	27.66	42.6 %
50.1	2.16	194° F.	21.75	29.14	33.9 %
50.2	1.44	129° F.	26.48	33.54	26.6 %

Both series of experiments show great losses by cylinder condensation ; they show also that these losses increase with the rate of expansion ; and they show greater losses with marine than with land boilers. This effect of cylinder condensation and wet steam can also be partially counteracted by steam or hot air jackets around the cylinders.

In his admirable work on the Steam Engine, Mr. D. K. Clark gives a number of carefully prepared tables on the Practice of Expansive Working in Steam Engines. By comparing in these the amount of steam shown by the indicator cards on the basis of dry saturated steam with the actual feed water used, we find the percentage of loss due to cylinder condensation and entrainment. This is figured in percentages of the calculated amounts, and *therefore shows how much should be added to estimates based on indicator cards to find the actual evaporation necessary for a required amount of work in a given engine.* The H. P. of the engine, the total initial pressure above vacuum in the cylinder, the total rate of expansion, and the superheat are given, as the figures can only be used under similar conditions.

TABLE NO. 46.

Table Illustrating Cylinder Condensation and Entrainment.

E. D. M.

KIND OF ENGINE.	H. P. of Engine.	Total Initial Press- ure, Pounds.	Superheat.	Total rate of Ex- pansion.	Pounds Water p. H. P. p. hour. Calculated fr'm In- dicator Diagram.	Actual Weight.	Differ'ce. Per Cent. Cylinder Condensa- tion & en- trainment.
Porter-Allen, not jacketed...........	66.	76	35.5° F.	6.34	24.69	25.81	4.5%
Cornish, steam jacketed.............	124.6	34	None.	3.62	16.76	20.72	19.2%
" " "	149.5	27	"	2.81	18.59	21.38	13.0%
" " "	190.7	36	"	3.66	15.82	18.82	21.4%
" " "	217.0	40	"	3.65	16.70	20.08	16.9%
Reynolds Corliss, not jacketed,cond'g	165.0	101	"	6.83	16.67	20.37	18.2%
" " non-condensing....	138.7	105	"	5.57	21.49	23.07	6.8%
Harris-Corliss, no jacket, condensing	167.4	105	"	7.39	15.83	19.15	17.3%
" " " non-condensing	135.8	104	"	6.55	21.02	23.68	11.2%
Wheelock, " condensing	160.4	1t3	"	6.64	15.26	19.22	20.6%
" " " non-condensing	141.8	103	"	5.23	20.80	24.61	15.4%
Corliss, steam jacket, condensing....	418.3	35	"	4.69	14.51	17.4	20.0%
Hirn, no jacket, condensing.........	142.4	60	150°F.	3.75	16.42	17.2	4.4%
" " "	134.6	54	None.	3.75	18.14	22.41	19.1%
" " "	111.6	56	85°F	5.84	14.20	16.16	12.1%
" " "	106.3	55	None.	5.84	14.42	19.93	27.6%
Cornish, steam jacket, condensing...	101.8	33	"	6.85	22.06	22.94	4.0%
Woolf Comp'd Cond'g, steam jacket	46.2	51	"	11.59	17.62	22.32	26.6%
" " " no jacket....	27.8	48	"	14.73	21.44	32.72	52.6%
" " Pump'g En, st'm jckt	118.4	36	"	11.35	17.19	22.62	31.6%
Woolf Comp'd Marine, steam in jckt.	96.3	90.6	"	7.68	16.15	21.72	34.5%
" " " no jacket....	72.9	89.7	"	7.15	15.48	23.34	50.7%
Same Eng, 2d cyl. only, steam jacket	78.0	78.2	"	9.49	18.08	27.09	49.8%
" " " no jacket....	69.4	89.0	"	8.25	18.71	30 32	62.0%
Receiver Comp'd Marine, steam jckt.	217.6	66.2	"	5.12	18.44	20.24	9.8%
Marine Cond'g, 1 cylinder, no jacket	201.1	67.6	"	3.17	18.02	26.53	47.2%
" " "	204.7	45.1	"	3.47	20.79	28.09	35.1%
Single Cylinder, ⌠no jacket..........	88.7	24.5	"	1.76	27.66	42.27	52.8%
American, ⎮steam in jacket...	96.5	25.2	"	1.75	29.27	37.34	20.8%
Marine, ⎬no jacket..........	185.8	53.2	"	3.83	19.24	25.93	34.7%
Condensing, ⎮steam in jacket...	171.8	53.5	"	4.01	18.27	21.86	19.6%
Condensing, ⎮no jacket..........	249.5	79.2	"	5.41	16.95	23.80	40.5%
Condensing, ⌡steam in jacket...	283.1	82.3	"	5.19	16.88	21.12	25.1%

150 H. P. Heine Boiler Ready for Shipment.

We see then that a calculation of water consumption from indicator cards may be anywhere from 4 per cent. to 62 per cent. out of the way.

We note further that superheating may counteract on the average all but 7 per cent. of the loss by moisture; careful lagging and good boilers may reduce it to 11.2 per cent. in the best of non-condensing engines; steam jackets in condensing engines may limit it to an average of 22.5 per cent., while in unjacketed condensing engines we may expect an average of 36.8 per cent.

Here again the land boilers show their advantage over the marine types. The average loss in steam jacketed land engines is 19.46 per cent. against 26.6 per cent. for the same type of marine engines; without jackets the land practice shows 21 per cent. loss against 46.1 per cent. for marine. It is evident that this discrepancy is in the boilers, and not in the engines, since marine engines are even more carefully built than land engines.

In specifying horizontal tubular or return tubular boilers for their work, careful engineers insist that the steam shall contain not more than 2 per cent. (sometimes 3 per cent.) of entrained water. This is considered good work for that type of boiler, and ample heating surface, and large liberating area and steam space are necessary to attain it.

Well designed water tube boilers give much better results. Several well authenticated tests of Heine Safety Boilers record entrainments as low as 1-8 of 1 per cent., and 1-2 of 1 per cent. when forcing 50 per cent. above rating, and from 1-12 of 1 per cent. entrainment to 1-7 of 1 per cent. super-heat at rating. Here then is a chance for economy in the engine gained by the boiler in addition to its own economy in fuel.   E. D. M.

The Rating of Boilers.

R. H. T.

It is considered usually advisable to assume a set of practically attaina-ble conditions in average good practice, and to take the power so obtainable as the measure of the power of the boiler in commercial and engineering transactions. The unit generally assumed has been usually the weight of steam demanded per horse power per hour by a fairly good steam engine. This magnitude has been gradually decreasing from the earliest period of the history of the steam engine. In the time of Watt, one cubic foot of water per hour was thought fair; at the middle of the present century, ten pounds of coal was a usual figure, and five pounds, commonly equivalent to about forty pounds of feed water evaporated, was allowed the best engines. After the introduction of the modern forms of engine, this last figure was reduced 25 per cent., and the most recent improvements have still further lessened the consumption of fuel and of steam. By general consent the unit has now become thirty pounds of dry steam per horse power per hour, which repre-sents the performance of good non-condensing mill engines. Large engines, with condensers and compounded cylinders, will do still better. A committee of the American Society of Mechanical Engineers recommended thirty pounds as the unit of boiler power, and this is now generally accepted. They advised that the commercial horse power be taken as *an evaporation of 30 pounds of water per hour from a feed water temperature of 100° Fahrenheit into steam at 70 pounds gauge pressure*, which may be considered to be equal to 34½ units of evaporation, that is, to 34½ pounds of water evaporated from a feed water temperature of 212° Fahrenheit into steam at the same temperature. This standard is equal to 33,305 British thermal units per hour.

It was the opinion of this committee that a boiler rated at any stated power should be capable of developing that power with easy firing, moderate draught, and ordinary fuel, while exhibiting good economy, and at least one-third more than its rated power to meet emergencies.

Kansas City Water Works,
KANSAS CITY, MO.
Contains 800 H. P. Heine Boilers.

The Energy Stored in Steam Boilers.

R. H. T.

A steam boiler is not only an apparatus by means of which the potential energy of chemical affinity is rendered actual and available, but it is also a storage reservoir, or a magazine, in which a quantity of such energy is temporarily held; and this quantity, always enormous, is directly proportional to the weight of water and of steam which the boiler at the time contains. The energy of gunpowder is somewhat variable, but a cubic foot of heated water under a pressure of 60 or 70 lbs. per square inch has about the same energy as one pound of gunpowder. At a low red heat water has about 40 times this amount of energy. Following are presented the weights of steam and of water contained in each of the more common forms of steam boilers, the total and relative amounts of energy confined in each under the usual conditions of working in every day practice, and their relative destructive power in case of explosion:

Broad Street Station of the Pennsylvania R. R. Co.,
PHILADELPHIA, PA.
Contains 2000 H. P. Heine Boilers.

TABLE NO. 47.

Total Stored Energy of Steam Boilers.

TYPE	AREA OF		Pressure, Pounds Per Square Inch	Rated Power, H.P.	WEIGHT OF (Pounds)			AVAILABLE STORED ENERGY IN (Foot Pounds)			ENERGY PER LB. OF (Foot lbs.)		MAXIMUM HEIGHT OF PROJECTION.* (Feet)		INITIAL VELOCITY. (Ft. Per Sec.)	
	Grate Surf.	Heat Surf.			Boiler.	Water.	Steam.	Water.	Steam.	Total.	Boiler.	Total Wt.	Boiler.	Total.	Boiler.	Total.
	Square Feet.															
1. Plain Cylinder	15	120	100	10	2500	5764	11.325	46605200	676698	47281898	18913	5714	18913	5714	1103	606
2. Cornish	36	730	30	60	16950	27471	31.45	57570750	709310	58260060	3431	1314	3431	1314	471	290
3. Two-flue Cylinder	20	400	150	35	6775	6840	37.04	80572050	2377357	82949407	12243	6076	12243	6076	888	625
4. Plain Tubular	30	852	75	60	9500	8255	20.84	50008790	1022731	51031521	5372	2871	5372	2871	588	430
5. Locomotive	22	1070	125	525	19400	5260	21.67	52561075	1483896	54044971	2786	2189	2786	2189	423	375
6. Locomotive	30	1350	125	650	25000	6920	31.19	69148790	2135802	71284592	2851	2231	2851	2231	428	379
7. Locomotive	20	1200	125	600	20585	6450	25.65	64452270	1766447	66218717	3219	2448	3219	2448	455	397
8. Locomotive	15	875	125	425	14020	6330	19.02	64253160	1302431	65555591	4677	3213	4677	3213	549	455
9. Scotch Marine	32	768	75	300	27045	11765	29.80	71272370	1462430	72734800	2689	1873	2689	1873	416	348
10. Scotch Marine	50.5	1119.5	75	350	37972	17730	47.20	107408340	2316392	109724732	2889	1968	2889	1968	431	356
11. Flue and Return Tubular	72.5	2324	30	200	56000	42845	69.81	90531490	1570517	92101987	1644	931	1644	931	325	245
12. Flue and Return Tubular	72	1755	30	180	56000	48570	73.07	102628410	1643854	104272264	1862	996	1862	996	346	253
13. Water Tube	70	2806	100	250	34450	21325	35.31	172455270	2108110	174563380	5067	3073	5067	3073	571	445
14. Water Tube	100	3000	100	250	45000	28115	58.50	227366000	3513830	230879830	5130	3155	5130	3155	575	450
15. Water Tube	100	3000	100	250	54000	13410	23.64	108346670	1311377	109624283	2030	1626	2030	1626	361	323

*This means the height to which the boiler, or boiler and contents, would be thrown if it went up in one piece, and hence all in one direction.

E. D. M.

The stored available energy in *water-tube boilers* is usually less than that of any of the other stationary boilers, and not very far from the amount stored, pound for pound, by the plain tubular boiler, the best of the older forms. It is evident that *their admitted safety from destructive explosion* does not come from this relation, however, but from the division of the contents into small portions, and especially from those details of construction which make it tolerably certain that any rupture shall be local. A violent explosion can only come of the general disruption of a boiler and the liberation at once of large masses of steam and water.

The Mallinckrodt Building, St. Louis, Mo.,
Contains 300 H. P. Heine Boilers.

Heating Buildings by Steam.

In heating buildings by steam we have two things to consider. First, the amount of fresh air entering the building per hour which has to be heated from the external to the desired internal temperature, and second, the amount of heat to be supplied to take the place of what is lost by conduction through walls, windows, roofs, ceilings and doors and thence by radiation and convection to the outer air.

It is generally customary to assume the air to be warmed as entering the house at $0°$ F., and in the United States the rule is to require an interior temperature of $70°$ F. The weight of 1 cu. ft. of air at $0°$ F. is 0.086 lbs ; its specific heat at constant pressure is 0.2377 (see Table No. 7). Therefore, to raise 1 cu. ft. of air at $0°$ F. one degree in temperature, we require $0.086 \times 0.2377 = 0.02$ H. U. To bring it from $0°$ to $70°$ will take 1.4 H. U. This of course is true only when the air is measured at the inlet opening ; for as it grows warmer it expands and a cu. ft. weighs less.

The amount of heat required to replace that dissipated through the exposed surfaces of the building can be figured from the following diagram, Table No. 48, which has been prepared by Mr. Alfred R. Wolff, M. E. It is "the graphical interpretation, in American units, of the practice and coefficients prescribed by law by the German Government in the design of the heating plants of its public buildings, and generally used in Germany for all buildings." Mr. Wolff has checked the coefficients by examples of good American practice, and found satisfactory agreement in the results.

Single Skylight
Window
Very Hight Window
Double Skylight
Double Window
Door
Door (hollow, heated)
Ceiled
Air

8"
12"
16"
20"
24"
28"
32"
36"
40"

110
100
90
80
70
60
50
40
30
20
10
0

$K(t - t_o)$

Thickness of Wall

10° 20° 30° 40° 50° 60° 70° 80° 90° 100°   120°   140°   160°   180°

Fahrenheit.

$(t - t_o)$

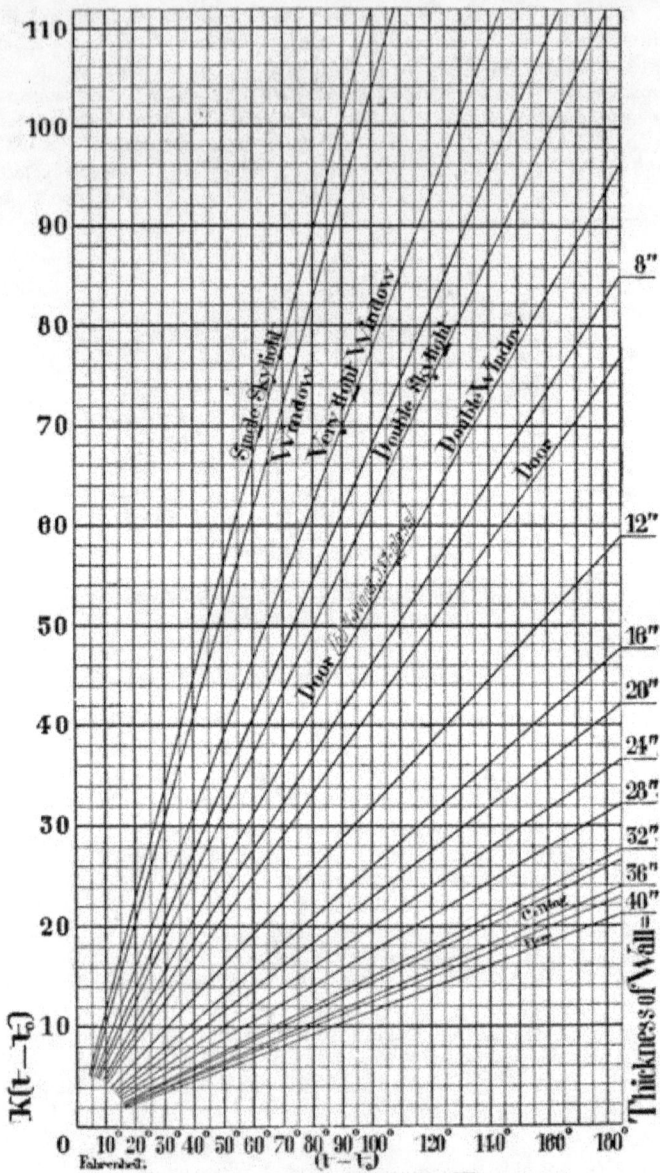

Heat Transmitted, in British Thermal
Units, per Square Foot of Surface, per Hour.

The formula for the loss is $Q = S \times K \times (t - t_0)$.

K is the loss by transmission in B. T. U. per hour per square foot of outer surface, per degree F. difference in temperature on the two sides.

S the number of square feet of transmitting surface, t the interior, and t_0 the exterior temperature in degrees Fahrenheit, of the air.

The values of K are given in the following table·

TABLE NO. 49

A. R. W.

For each square foot of brick wall of thickness:

Thickness of brick wall=	4″	8″	12″	16″	20″	24″	28″	32″	36″	40″
K =	0.68	0.46	0.32	0.26	0.23	0.20	0.174	0.15	0.129	0.115

1 square foot, wooden beam construction, as flooring, $K = 0.083$
    planked over, or ceiled: as ceiling, $K = 0.104$
1 square foot, fireproof construction, as flooring, $K = 0.124$
    floored over: as ceiling, $K = 0.145$
1 square foot, single window _____ $K = 0.776$
1 square foot, single skylight _____ $K = 1.118$
1 square foot, double window _____ $K = 0.518$
1 square foot, double skylight_____ $K = 0.621$
1 square foot, door _____ $K = 0.414$

These coefficients are to be increased respectively, as follows:

Ten per cent. where the exposure is a northerly one and winds are to be counted on as important factors.

Ten per cent. when the building is heated during the daytime only, and the location of the building is not an exposed one.

Thirty per cent. when the building is heated during the daytime only, and the location of the building is exposed.

Fifty per cent. when the building is heated during the winter months intermittently, with long intervals (say days or weeks) of non-heating.

In using this table it is necessary to know the conditions as to temperature of adjoining buildings having the same party-wall and of the different stories, cellar, attic, etc., of the building to be heated. Then with the plans of the building at hand the total square feet of each kind of surface can be measured and the estimate rapidly made from the diagram, Table No. 48, as follows:

Find the difference in temperatures $t - t_0$ on the lower horizontal line; run up the vertical line thus found until it intersects the diagonal line representing the kind of surface; follow the horizontal line to the left and read on the vertical scale the value of K $(t - t_0)$.

F. i., 70° required in the room, temperature of adjoining hallway being 10°. Find difference 60°. The division wall being 24″; run up on the 60° line to the diagonal for 24″ wall, then follow the horizontal line to the left and you find 12 H. U. as the value of K $(t - t_0)$. Suppose there is a door in the wall; the 60° line strikes it midway between 24 and 26 on the vertical scale, hence we have 25 H. U. for every square foot of door.

— 85 —

Pulaski Iron Works,
PULASKI CITY, VA.
Contains 1600 H. P. Heine Boilers.

For the amount of air which should be admitted to each room, Morin gives

Cubic feet of air required for ventilation per head per hour.

Hospitals, ordinary maladies 2470
Hospitals, wounded, etc. 3530
Hospitals, in times of epidemic 5300
Theatres 1585
Assembly rooms, prolonged sittings 2120
Prisons 1760
Workshops, ordinary 2120
Workshops, insalubrious conditions 3530
Barracks, day 1060, at night 1760
Infant schools 706
Adult schools 1410
Stables 7060

Having determined the total number of H. U. required for each room, the kind and quantity of the radiating surface is next to be determined.

The character of the surfaces determines their efficiency.

Mr. P. Kaeuffer, M. E., of Mayence, Germany, has made a number of careful experiments on radiating surfaces, the results of which, recalculated for American units, we give in

TABLE NO. 31.

Transmission of heat by radiating surfaces, per square foot per hour in B.T.U.

Smooth vertical plane 406
Vertical plane with about 80% surface in ribs or corrugations 170
Smooth vertical pipe surface 480
Vertical tube with 67% of surface in corrugations 221
Horizontal smooth tube or pipe 369
Horizontal tube with 67% of surface in corrugations 185

NOTE.—This table is correct for steam of 15 to 22 pounds pressure ; for exhaust steam reduce in proportion to temperature, except for corrugated and ribbed surfaces, which lose very rapidly for low steam temperatures. For hot water, 50 per cent. of the tabular numbers are approximately correct.

Approximately (for St. Louis conditions) 9 feet of 1″ pipe with exhaust steam, or 6 feet of 1″ pipe with 50 pounds steam, will heat 1000 cubic feet of air 70° per hour.

French practice is about 1 square foot of radiating surface for 230 cubic feet of space for exhaust steam. This is about 13 feet run of 1″ pipe for 1000 cubic feet of space.

Mr. Wolff gives 250 H. U. per hour per square foot surface for ordinary bronzed cast iron radiators, and 400 H. U. for non-painted radiating surfaces, counting steam pressure from 3 to 5 pounds per square inch. (About 60% of these amounts for hot water heating.) When the *total number of heat units required* are known the *work of the boiler* can be directly estimated from them ; bearing in mind that if the water condensed in the radiators is returned to the boiler at 212°, we have in each pound of exhaust steam 965.8 heat units available, in steam of 2 pounds, 5 pounds, or 10 pounds gauge pressure, we have 967.5 H. U., 969.7 H. U., or 974.1 H. U. respectively per pound of steam delivered to the system.

As we have seen by Table No. 51, the effectiveness of radiating surfaces varies too much to make it the basis of the amount of boiler power required. Still, for rough approximations it is so used; some experts estimate a square foot of boiler-heating surface for every 7 or 10 square feet of radiating surface; some go as far as 1 to 15. Mr. Kaeuffer's estimates are for about 1 square foot of boiler H. S. for 6 square feet of the best and 18 square feet of the poorest radiating surface. (See Table 51.) In roughly estimating from the cubical contents of buildings, we must observe that small buildings, having proportionately more exposed wall and window surface per 1000 cubic feet of contents, require proportionately more boiler power. And as the amount of ventilation necessary depends on the nature of the use of the building, this also affects the amount of boiler power required.

TABLE NO. 52.

Approximate Number of Cubic Feet which 1 H. P. in Boiler will Heat.

Hospitals, exposition buildings, etc., with much window
surface_____ 6000 to 8000
Dwellings, stores, small shops, etc_____ 8000 to 12000
Foundries, large workshops, etc_____ 8000 to 16000
Theaters, schools, prisons, churches, etc_____10000 to 18000
Armories, gymnasiums, etc_____15000 to 25000

The remarks about increase in the value of K under Table No. 49 apply directly to increase in boiler power for similar conditions.

Heating Liquids by Steam.

Liquids may be heated by blowing the steam into them through a number of small openings, or by passing the steam through a coil of pipe submerged in the liquid, or by passing the steam through an external casing. In the former case dilution results, and any impurities in the steam of course enter into and foul the liquid. The latter two methods are therefore more frequently adopted in practice. In heating water, it is found that the work done per unit of surface and temperature is greatly increased when boiling begins and evaporation takes place, even though the difference in temperature be less. In this connection the experiments of Thos. Craddock are interesting. A velocity of 3 feet per second of the water doubled the rate of transmission in still water; he found that this circulation became more valuable as the difference in temperatures became less.

The following table by Mr. Thos. Box illustrates this point. When evaporation had set in and caused *circulation, the effectiveness of the surfaces was trebled*, although the difference of temperature was only one-third of that in the still water, an apparent nine-fold increase.

TABLE NO. 53.

Table of Experiments on the Power of Steam Cased Vessels and Steam Pipes in Heating Water.

Box.

Temperature of the water heated.			Temp. of the Steam.	Difference of Temperature of Steam and Water.	Units per sq. ft. per hr. for 1° difference of temp.				Kind of Heater.
					By Experiment.		By Table.		
Mini-mum.	Maxi-mum.	Mean.			Units.	Mean.	Units.	Mean.	
Deg.	Deg.	Deg.	Deg.	Deg. Deg.					
65	110	......	212	147 to 202	230		216		Vertical tube.
60	102½	......	212	152 to 109½	207	216	210	216	Vertical tube.
69	109½	......	212	143 to 102½	210		221		Vertical tube.
39	212	......	274	235 to 62	335		325		Steam cased vessel.
46	212	......	274	228 to 62	315	325	333	329	Worm.
*......	......	212	274	62	974		1000		Worm.
*......	......	212	250	38	1020	997	1000	1000	Worm.

*NOTE—These two results were evaporation of water already at 212° F., the preceding one showing that only about one-third as much heat was transmitted in heating still water.

A remarkable fact was noted in some experiments in this line by Mr. B. G. Nichol, in 1875, namely, that a horizontal position of the pipe was more effective than a vertical one. This is the reverse of what is found in heating air. (Compare Table No. 51, Kaeuffer.)

Safety Valves.

It was formerly the custom to proportion the Safety Valves according to the heating surface. But as the performance per square foot of H. S. varies widely in different boilers (from 2 to 15 lbs. hourly evaporation), the wiser plan of giving the safety valves a fixed ratio to the grate area has been adopted.

The United States Treasury Department, through its Board of Supervising Inspectors of Steam Vessels has established the following rules:

"Lever safety valves to be attached to marine boilers shall have an area of not less than *one square inch to two square feet* of grate surface in the boiler, and the seats of all such safety valves shall have an angle of inclination of 45° to the center line of their axes.

"The valves shall be so arranged that each boiler shall have one separate safety valve, unless the arrangement is such as to preclude the possibility of shutting off the communication of any boiler with the safety valve, or valves employed. This arrangement shall also apply to lock-up safety valves when they are employed.

"Any spring-loaded safety valves constructed so as to give an increased lift by the operation of steam, after being raised from their seats, or any spring-loaded safety valve constructed in any other manner, or so as to give an effective area equal to that of the afore-mentioned spring-loaded safety valve, may be used in lieu of the common lever-weighted valves on all boilers on steam vessels, and all such spring-loaded safety valves shall be required to have an area of not less than one square inch to three square feet of grate surface of the boiler, and each spring-loaded safety valve shall

1500 H. P. Plant of Heine Boilers, part of 4,500 H. P. Plants of Broadway and 7th Ave. Cable Ry. Co., NEW YORK.

be supplied with a lever that will raise the valve from its seat a distance of not less than that equal to one-eighth the diameter of the valve opening, and the seats of all such safety valves shall have an angle of inclination to the center line of their axis of 45'. But in no case shall any spring-loaded safety valve be used in lieu of the lever-weighted safety valve without having first been approved by the Board of Supervising Inspectors."

This rule, so far as it applies to lever-weighted safety valves, is identical with the Board of Trade Rule of Great Britain.

It has, however, the one defect that it takes no account of the pressure carried. And a safety valve of correct size for 50 lbs. pressure would be more than three times too large for 200 lbs. pressure, and may become a source of danger.

The PHILADELPHIA BOILER LAW takes this into account and orders that the "least aggregate area of safety valve (being the least sectional area for the discharge of steam) to be placed upon all stationary boilers with natural or chimney draft, may be expressed by the formula

$$A = \frac{22.5\,G}{P + 8.62}$$

in which A is the area of combined safety valves in inches. G is area of grate in square feet. P is pressure of steam in pounds per square inch to be carried in the boiler above the atmosphere. The following table gives the results of the formula for one square foot of grate as applied to boilers used at different pressures.

TABLE NO. 54.

Pressure per Square Inch.

10	20	30	40	50	60	70	80	90	100	110	120	150	175
1.21	0.79	0.58	0.46	0.38	0.33	0.29	0.25	0.23	0.21	0 19	0.17	0.142	0.123

Valve area in square inches, corresponding to one square foot of grate.

Horse-Power and Steam Consumption of Pumping Engines.

Multiply the number of million gallons pumped per 24 hours by the total head (including suction head), expressed either in feet or in pounds. This product multiplied by 0.176 if the head is stated in feet, or by 0.405 if the head is given in pounds, will be the horse-power of work done by the water end, or the horse-power of the water column. Thus f. i., a 15 million gallon engine with 260 ft. total head does $15 \times 260 \times 0.176 = 686.4$ horse-power; and a 15 million gallon engine raising water against a total pressure of 110 lbs. does $15 \times 110 \times 0.405 = 668.3$ horse-power. It is the universal practice among engineers to express the economic efficiency of a pumping engine by what is called its "*duty*," *i. e.* the number of millions of foot pounds of work it will do for every hundred pounds of coal burned under the boilers.

Generally specifications base the duty to be guaranteed on an assumed evaporation of 10:1 or state that for every 1000 lbs. of steam (measured by the boiler feed-water) such duty is to be given.

Either method fails to define where the duty of the boiler ends and that of the engine begins, since neither states from what temperature of feed to what pressure of steam the boilers are to evaporate.

By the established practice among mechanical engineers, boiler performances are compared as to economy on the basis of evaporation from and at 212° F. In the absence of any specific statement the assumed evaporation of 10 to 1 would, therefore, be thus construed, and as this is about the best performance that can be safely counted on per pound of best coal, it virtually becomes the basis of calculation.

A pumping engine of 100 million duty will require 19.8 lbs. feed-water per hour per horse-power of work in water column, based on an evaporation of 10 lbs. water per pound of coal from and at 212° F.

But as pumping engines are constructed for steam pressures varying from 75 lbs. for high pressure single cylinder engines to 175 pounds for triple expansion ; and as the feed-water may be, say 100° F. the temperature of the hot well, or 212° F. from a good exhaust heater, the amount of feed-water required by the engine per horse-power per hour will vary according to these conditions.

The higher the steam pressure the greater the amount of energy available in each pound of steam. The lower the feed temperature the larger the proportion of the boiler's work which had to be expended in merely heating the water up to the boiling-point. On this basis the following table has been figured :

TABLE NO. 55.

Showing Lbs. Feed–Water per Horse-power required by Pumping Engines per Hour.

E. D. M.

Duty.	From Feed at 212° F. to Steam of:					From Feed at 100° F. to Steam of:					Equivalent to Boiler Work in U. of E., or Pounds from and at 212° F.
	75 lbs.	100 lbs.	125 lbs.	150 lbs.	175 lbs.	75 lbs.	100 lbs.	125 lbs.	150 lbs.	175 lbs.	
110 Mill.	17.37	17.30	17.23	17.16	17.09	15.64	15.57	15.50	15.44	15.38	18.00
100 Mill.	19.11	19.03	18.95	18.87	18.80	17.20	17.12	17.05	16.98	16.92	19.80
90 Mill.	21.23	21.14	21.06	20.97	20.88	19.11	19.02	18.94	18.87	18.80	22.00
80 Mill.	23.90	23.80	23.70	23.60	23.50	21.50	21.40	21.31	21.22	21.15	24.75
70 Mill.	27.30	27.19	27.07	26.96	26.86	24.57	24.46	24.36	24.26	24.17	28.29
60 Mill.	31.85	31.71	31.58	31.45	31.33	28.67	28.53	28.42	28.30	28.20	33.00
50 Mill.	38.22	38.06	37.90	37.74	37.60	34.40	34.24	34.10	33.96	33.84	39.60

NOTE. The horse-power is the H. P. of the water column. The evaporation is assumed at 10 lbs. water from and at 212° F. per lb. of coal.

Economy in boilers is always stated in "*pounds of water evaporated from and at 212° F. per pound of fuel,*" designated as "*Units of Evaporation.*" (See Vol. VI, Transactions Am. Soc. M. E.—1881).

Unless a contract specifically provides otherwise the "*assumed evaporation*" is to be so understood.

The last vertical column of the table gives the equivalent work for the boiler in each case per horse-power of the water column ; in fact, all the figures in each horizontal line are exact equivalents of each other. Again, comparing the vertical columns with each other it is clear that an engine provided with a first-class feed-water heater will save 11.1% over the same engine relying simply on its hot well.

Given an assumed evaporation per pound of such coal as the guarantee is based on ; or the evaporation found by actual test of the boilers. Divide the figure in the last vertical column by such evaporation, and you have the number of pounds of the coal per horse-power in each case.

E. D. M.

Condensers.

H. R. W.

When steam expands in the cylinder of a steam engine, its pressure gradually reduces, and ultimately becomes so small that it cannot profitably be used for driving the piston. At this stage a time has arrived when the attenuated vapor should be disposed of by some method, so as not to exert any back pressure or resistance to the return of the piston. If there were no atmospheric pressure, exhausting into the open air would effect the desired object. But, as there is in reality a pressure of about 14.7 pounds per square inch, due to the weight of the super-incumbent atmosphere, it follows that steam in a non-condensing engine cannot economically be expanded below this pressure, and must eventually be exhausted against the atmosphere, which exerts a back pressure to that extent.

It is evident that if this back pressure be removed, the engine will not only be aided, by the exhausting side of the piston being relieved of a resistance of 14.7 pounds per square inch, but moreover, as the exhaust or release of the steam from the engine cylinder will be against no pressure, the steam can be expanded in the cylinder quite, or nearly, to absolute 0 of pressure, and thus its full expansive power can be obtained.

Contact, in a closed vessel, with a spray of cold water or with one side of a series of tubes, on the other side of which cold water is circulating, deprives the steam of nearly all its latent heat, and condenses it. In either case the act of condensation is almost instantaneous. A change of state occurs, and the vapor steam is reduced to liquid water. As this water of condensation only occupies about one sixteen-hundredths of the space filled by the steam from which it was formed, it follows that the remainder of the space is void or vacant, and no pressure exists. Now, the expanded steam from the engine is conducted into this empty or vacuous space, and, as it meets with no resistance, the very limit of its usefulness is reached.

The vessel in which this condensation of steam takes place is the condensing chamber. The cold water that produces the condensation is the injection water; and the heated water, on leaving the condenser is the discharge water.

To make the action of the condensing apparatus continuous, the flow of the injection water, and the removal of the discharge water including the water from the liquifaction of the steam, must likewise be continuous.

The vacuum in the condenser is not quite perfect, because the cold injection water is heated by the steam, and emits a vapor of a tension due to the temperature. When the temperature is 110 degrees Fahrenheit, the tension or pressure of the vapor will be represented by about 4" of mercury ; that is, when the mercury in the ordinary barometer stands at 30", a barometer with the space above the mercury communicating with the condenser,

Cape Town Tramways Co., Limited.
CAPE TOWN, AFRICA.
900 H. P. of Heine Boilers.

will stand at about 26″. The imperfection of vacuum is not wholly traceable to the vapor in the condenser, but also to the presence of air, a small quantity of which enters with the injection water and with the steam; the larger part, however, comes through air leaks and faulty connections and badly packed stuffing boxes. The air would gradually accumulate until it destroyed the vacuum, if provision were not made to constantly withdraw it, together with the heated water, by means of a pump.

The amount of water required to thoroughly condense the steam from an engine is dependent upon two conditions: the total heat and volume of the steam, and the temperature of the injection water. The former represents the work to be done, and the latter the value of the water by whose cooling agency the work of condensation of the steam is to be accomplished. Generally stated, with 26″ vacuum, the injection water at ordinary temperature, not exceeding 70 degrees Fahrenheit, from 20 to 30 times the quantity of water evaporated in the boilers will be required for the complete liquifaction of the exhaust steam. The efficiency of injection water decreases very rapidly as its temperature increases, and at 80 degrees and 90 degrees Fahrenheit, very much larger quantities are to be employed. Under the conditions of common temperature of water and a vacuum of 26″ of mercury, the injection water necessary per H. P. developed by the engine, will be from 1¼ gallons per minute when the steam admission is for one-fourth of the stroke, up to 2 gallons per minute when the steam is carried three-fourths of the stroke of the engine.

The power exerted by a steam engine during a single stroke of a piston, is due directly to the difference between the pressures on the opposite sides of the piston. Newton said, "all force is *vis a tergo;*"—a push from behind. A vacuum does not in itself give power. It only effects a removal of resistance from the retreating side of the piston, and consequently adds just so much activeness to the other, or pushing side. The value of a vacuum of 26″ of mercury to an engine, may be generally approximated by considering it to be equivalent to a net gain of 12 lbs. average pressure per square inch of piston area. It is obvious that this amount of power gained bears nearly the same ratio to the power developed by the engine when non-condensing, as 12 lbs. does to the mean effective, or average pressure of the steam in the cylinder. So, if the mean effective pressure is known, a close idea of the percentage of gain that will be derived by the use of a vacuum with a non-condensing engine, may be arrived at.

By the use of Watt's formula, in which,

A = Area of piston in square inches.

V = Velocity of piston in feet per minute.

M. E. P. = Mean effective pressure of the steam in pounds per square inch on the piston.

$$\frac{A \times V \times M.E.P.}{33000} = \text{Horse Power.}$$

And by substituting 12 for M. E. P., the value of vacuum of 12 lbs. expressed in horse power is found.

$$\frac{A \times V \times 12}{33000} = \text{Horse power made available by vacuum.}$$

Table of Mean Effective Pressures.

The following graphical table will afford a ready and comprehensive means of ascertaining the mean effective pressure of steam in an engine cylinder when the initial steam pressure and point of cut-off, or the number of expansions of the steam, are known.

It should be borne in mind that "absolute pressure" is calculated from the absolute vacuum of the barometer, while "gauge pressure" as indicated by the ordinary pressure gauge, begins with atmospheric pressure as its zero; consequently "absolute pressure" is nearly 15 pounds greater than "gauge pressure."

TABLE NO. 56.

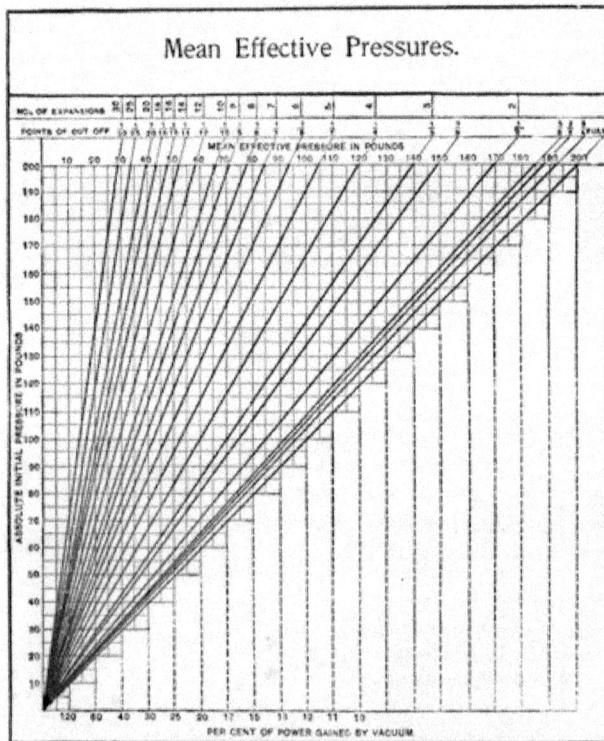

Mean Effective Pressures.

(From Special Catalogue of The Worthington Condenser.)

The left hand vertical column of figures gives the initial (absolute) steam pressure, and the upper horizontal row, the number of expansions that correspond to the several points of cut-off; directly under this is a similar one of the mean effective pressures.

To determine the M. E. P. produced in an engine cylinder with an initial pressure of 90 pounds steam (gauge pressure), cut-off at one-quarter stroke, expanded and finally exhausted into a vacuum; add 15 to 90, and find 105 in the initial pressure column; follow the horizontal line to the right until it intersects the oblique line which corresponds to ¼ cut-off. Then read the M. E. P. from the row of figures directly above, which in this case is 63 pounds.

If, as in a non-condensing engine, the steam is exhausted against atmospheric pressure, this 63 pounds M. E. P. should be reduced by 15 pounds, giving 48 pounds as the net result.*

By glancing down and reading on the lower scale the figures directly under the 48 pounds M. E. P. on the upper row, will be seen the percentage of power that a vacuum will add to an engine using 90 pounds "gauge pressure" steam, cut-off at one-quarter stroke. Thus, in this instance, the value of the vacuum is found to be between 25 and 30 per cent of the power of the engine when running non-condensing.

H. R. W.

---

*NOTE.—In condensing engines it will be safer to deduct from 3 to 5 pounds for imperfect vacuum, etc., and in non-condensing engines 16 to 18 pounds in place of 15 for back pressure, etc.

E. D. M.

Minneapolis Industrial Exposition Building, Minneapolis, Minn.,
With Heine Boiler Plant of 1000 H. P.

NOTE ON BOILER TESTS :—Table No. 57 gives the results of thirty-three tests made with various coals. To justly estimate the efficiency of the boiler from same, compare the heat values of the coals as given in Table 12.

Table No. 57. Results of Tests of Heine Boilers.

Number	Where Made	Observers' Name	Coal Used	Coal from	Rated Capacity	Duration of test, hours	Steam Pressure	Draft Pressure	Temp. of feed	Lbs. of water evap. per lb. coal, from and at 212°	Lbs. of water evap. per sq. ft. H.S. per hour, from and at 212°	Lbs. of coal per sq. ft. grate, per hour	Lbs. of coal per H.P. per hour	H.P. actually developed	Per cent over-rated capacity	Sq. ft. H.S. per H.P.	Per cent of entrainment
1	Chicago Athletic Ass'n	T. H. Nelson	Big Muddy	Ill	150	8	103.4	1.0	82.	8.41	8.974	42.9	3.68	326.	117.0	3.45	1/5 of 1%
2	Chicago Edison Co	T. H. Nelson	Big Muddy	Ill	366	8	114.3	1.08	72.36	8.58	6.193	37.0	3.61	353.1	51.5	5.73	1/10 of 1%
3	Edison Illuminating Co., St. Louis	F. G. Schlosser	Carterville	Ill	375	10	121.7	.75	66	8.33	5.403	2.4	3.72	476.	26.8	5.73	1/4 of 1%
4	Laclede Electric Light Co., St. Louis	J. F. Shulte	Nut	Ill	314	10	150.88	.8	200.	7.65	6.003	6.	4.04	445.	41.7	5.16	1/10 of 1%
5	Anheuser-Busch Brewery, St. Louis	C. E. Jones	Lump	Ill	300	8	96.76	1.4	89.6	6.88	7.225	4.394	4.49	536.	78.7	4.2	1/10 of 1%
6	N. K. Fairbanks, St. Louis	C. E. Jones	Vulcan Nut	Ill	300	8	81.5	.8	75.0	7.66	6.774	0.6	4.66	502.3	67.4	4.60	1/10 of 1%
7	Cupples Building, St. Louis	W. H. Bryan	Belleville	Ill	370	9	86.57	.87	210.3	7.58	4.593	4.9	4.64	14.9	12.1	6.63	
8	Mallinckrodt Building, St. Louis	W. H. Bryan	Gillespie	Ill	150	6	70.5	.6	185.1	7.36	4.862	7.6	4.21	382.1	22.1	6.38	Dry.
9	Mallinckrodt Chem. Works	W. B. Potter	Collinsville	Ill	375	10	91.1	.62	173.6	7.80	5.073	6.2	3.97	453.2	29.5	6.11	1/10 of 1%
10	Blish Milling Co., Seymour, Ind.	J. C. Brown	Lump and Slack	Ind	225	8	114.5	.4	194.16	7.62	4.603	1.3	254.0	246.	9.3	6.3	Dry.
11	Texarkana Ice Co., Texas	Robt. Hays	Lump and Slack	Ind. N	70	10	92.5	.6	185.	10.38	6.861	7.1	2.99	129.3	71.8	4.51	1/10 of 1%
12	Texas Star Flour Mill, Galveston	W. J. Green	Webster	Ind. N	200	9	121.	.6	104.3	8.36	5.322	2.163	1.3	287.71	43.8	3.9	
13	Cedar Rapids (Ia.) E. L. & P. Co.	J. A. Wathen	Slack	Iowa	120	5	116.8	.47	183.8	5.44	4.603	4.886	63	145.5	21.9	3.9	1/10 of 1%
14	Wathen Bros. & Co., Louisville	W. H. Neal	Jellico	Ky	300	6	70.2	.8	116.4	9.71	7.623	4.253	20384.23		94.7	3.9	1/10 of 1%
15	Armour Packing Co., Kansas City	G. E. Worthen	Lump	Mo	300	6	80.	.8	38.0	7.50	6.453	3.544	13478.4		59.5	4.81	1/10 of 1%
16	Kansas City Water Works	G. H. Barrus	Cumberland	Md	370	13	104.36	.22	131.0	10.91	5.652	4.772	48435.48		23.0	6.3	Dry.
17	James Row & Co., Troy, N. Y.	G. H. Barrus	Cumberland	Md	180	10	91.1	.90	90.8	9.83	5.701	4.343	15194.8		8.205	43	Dry.
18	Edison Illuminating Co., Boston	E. A. Hammond	Cumberland & Anthracite	Md	300	10	100.5	.6	204.5	10.34	4.361	6.443	00323.87		7.607.1		1/10 of 1%
19	James Roy & Co., Troy, N. Y.	G. H. Barrus	Anthracite	Pa	180	11	72.9	.23	97.1	9.14	5.381	4.553	40183.5		1.045.77		
20	Orrs & Co., Troy, N. Y.	P. H. Baerman	Anthracite D. & H.	Pa	115	12	66.3	3.5	32.0	9.79	5.941	7.723	16169.5		47.395.2		1/10 of 1%
21	Union Ice Co., Pittsburg, Pa.	M. Irvin	Pittsburgh	Pa	300	4	124.77	.89	40.0	8.29	7.981	4.763	78591.7		97.223.88		1/10 of 1%
22	Union Tobacco Works, Louisville	S. P. Liner	Pittsburgh	Pa	70	6	54.47	.4	110.	9.68	6.672	2.183	29129.3		84.7	4.3	1/10 of 1%
23	Troy (N. Y.) Iron & Steel Co.	H. Shaw	Bit. Slack	Pa	125	21.5	62.27	1.0	103.	8.12	5.031	8.3	85155.7		24.5	6.2	Dry.
24	Chicago Edison Co., Adams St	R. W. Francis	Youghiogheny	Pa	366	5	111.6	1.0	70.	10.08	6.753	0.963	08547.7		49.6	4.6	1/10 of 1%
25	Memphis (Tenn.) Water Works	J. J. DeKinder	Youghiogheny	Pa	300	15	111.26	.8	151.8	10.51	4.712	3.132	93346.0		15.696.48		Dry.
26	Bohlen-Huse Ice Co., Memphis	E. Hayden	Youghiogheny	Pa	110	24	86.6	.55	69.21	10.82	5.762	1.292	86167.13		51.9	5.4	1/10 of 1%
27	City Elec. L. Co., Kalamazoo, Mich.	M. E. Cooley	Youghiogheny	Pa	150	5	123.6	.6	104.0	10.39	4.941	9.172	98179.2		10.5	6.3	1/10 of 1%
28	C. C. Washburn Flour M. Co., Minn's	Prof. W. A. Pike	Youghiogheny	Pa	666	5	134.6	.75	111.1	10.42	4.542	9.623	31715.0		7.2	7.6	1/10 of 1%
29	Jersey City & Bergen Electric Ry	J. J. DeKinder	Moshannon	Pa	250	8	120.76	.7	55.5	9.28	6.792	6.353	71464.4		81.7	5.07	1/10 of 1%
30	San Leandro Elec. Ry., San Francisco	D. Dorward	Carbon Hill	Wash	130	6	115.7	.6	127.68	8.20	4.853	1.433	74153.4		18.0	6.30	1/10 of 1%
31	Risdon Iron & Loco. Wks., San Fran.	D. Dorward	Carbon Hill	Wash	193	8	98.5	.57	115.0	8.9	4.142	5.9	3.51193.0		0.0	7.5	1/10 of 1%
32	Magog Print Work, Canada	J. La Point	Slack	Canada	150	5	55.4	.5	92.9	6.82	5.092	9.354	44185.0		23.0	6.0	1/10 of 1%
33	Spring Valley Water Works, Cal.	D. Dorward	Sidney	Australia	175	24	95.	.6	140.0	8.95	3.923	0.3	3.46183.70		4.957.9		1/10 of 1%

CODE OF RULES FOR BOILER TESTS.

Recommended by the American Society of Mechanical Engineers, November, 1884.

A Committee of the Society is now engaged on a new and more complete Code of Rules,
which is to suggest Standard Coals for different localities.

Starting and Stopping a Test.

A test should last at least ten hours of continuous running, and twenty-four hours whenever practicable. The conditions of the boiler and furnace in all respects should be, as nearly as possible, the same at the end as at the beginning of the test. The steam pressure should be the same, the water-level the same, the fire upon the grates should be the same in quantity and condition, and the walls, flues, etc., should be of the same temperature. To secure as near an approximation to exact uniformity as possible in conditions of the fire and in the temperatures of the walls and flues, the following method of starting and stopping a test should be adopted:

Standard Method.—Steam being raised to the working pressure, remove rapidly all the fire from the grate, close the damper, clean the ash-pit, and as quickly as possible, start a new fire with weighed wood and coal, noting the time of starting the test and the height of the water-level while the water is in a quiescent state, just before lighting the fire.

At the end of the test, remove the whole fire, clean the grates and ash-pit, and note the water-level when the water is in a quiescent state ; record the time of hauling the fire as the end of the test. The water-level should be as nearly as possible the same as at the beginning of the test. If it is not the same, a correction should be made by computation, and not by operating pump after test is completed. It will generally be necessary to regulate the discharge of steam from the boiler tested by means of the stop-valve for a time while fires are being hauled at the beginning and at the end of the test, in order to keep the steam pressure in the boiler at those times up to the average during the test.

Alternate Method.—Instead of the Standard Method above described, the following may be employed where local conditions render it necessary :

At the regular time for slicing and cleaning fires have them burned rather low, as is usual before cleaning, and then thoroughly cleaned ; note the amount of coal left on the grate as nearly as it can be estimated ; note the pressure of steam and the height of the water-level—which should be at the medium height to be carried throughout the test—at the same time ; and note this time as the time of starting the test. Fresh coal, which has been weighed, should now be fired. The ash-pits should be thoroughly cleaned at once after starting. Before the end of the test the fires should be burned low, just as before the start, and the fires cleaned in such a manner as to leave the same amount of fire, and in the same condition, on the grates as at the start. The water-level and steam pressure should be brought to the same point as at the start, and the time of the ending of the test should be noted just before fresh coal is fired.

Hotel Majestic,
NEW YORK CITY.

During the Test

Keep the Conditions Uniform.—The boiler should be run continuously, without stopping for meal-times or for rise or fall of pressure of steam due to change of demand for steam. The draught being adjusted to the rate of evaporation or combustion desired before the test is begun, it should be retained constant during the test by means of the damper.

If the boiler is not connected to the same steam-pipe with other boilers, an extra outlet for steam with valve in same should be provided, so that in case the pressure should rise to that at which the safety valve is set, it may be reduced to the desired point by opening the extra outlet, without checking the fires.

If the boiler is connected to a main steam-pipe with other boilers, the safety valve on the boiler being tested should be set a few pounds higher than those of the other boilers, so that in case of a rise in pressure the other boilers may blow off, and the pressure be reduced by closing their dampers, allowing the damper of the boiler being tested to remain open, and firing as usual.

All the conditions should be kept as nearly uniform as possible, such as force of draught, pressure of steam, and height of water. The time of cleaning the fires will depend upon the character of the fuel, the rapidity of combustion, and the kind of grates. When very good coal is used, and the combustion not too rapid, a ten-hour test may be run without any cleaning of the grates, other than just before the beginning and just before the end of the test. But in case the grates have to be cleaned during the test, the intervals between one cleaning and another should be uniform.

Keeping the Records.—The coal should be weighed and delivered to the firemen in equal portions, each sufficient for about one hour's run, and a fresh portion should not be delivered until the previous one has all been fired. The time required to consume each portion should be noted, the time being recorded at the instant of firing the first of each new portion. It is desirable that at the same time the amount of water fed into the boiler should be accurately noted and recorded, including the height of the water in the boiler, and the average pressure of steam and temperature of feed during the time. By thus recording the amount of water evaporated by successive portions of coal, the record of the test may be divided into several divisions, if desired, at the end of the test, to discover the degree of uniformity of combustion, evaporation, and economy at different stages of the test.

Priming Tests.—In all tests in which accuracy of results is important, calorimeter tests should be made of the percentage of moisture in the steam, or of the degree of super-heating. At least ten such tests should be made during the trial of the boiler, or so many as to reduce the probable average error to less than one per cent, and the final records of the boiler test corrected according to the average results of the calorimeter tests.

On account of the difficulty of securing accuracy in these tests the greatest care should be taken in the measurements of weights and temperatures. The thermometers should be accurate to within a tenth of one degree, and the scales on which the water is weighed to within one-hundreth of a pound.

Analyses of Gases. Measurement of Air Supply, Etc.

In tests for purposes of scientific research, in which the determination of all the variables entering into the test is desired, certain observations should be made which are in general not necessary in tests for commercial purposes. These are the measurement of the air supply, the determination of its contained moisture, the measurement and analysis of the flue gases, the determination of the amount of heat lost by radiation, of the amount of infiltration of air through the setting, the direct determination by calorimeter experiments of the absolute heating value of the fuel, and (by condensation of all the steam made by the boiler) of the total heat imparted to the water.

The analysis of the flue gases is an especially valuable method of determining the relative value of different methods of firing, or of different kinds of furnaces. In making these analyses great care should be taken to procure average samples, since the composition is apt to vary at different points of the flue, and the analyses should be intrusted only to a thoroughly competent chemist, who is provided with complete and accurate apparatus.

As the determination of the other variables mentioned above are not likely to be undertaken except by engineers of high scientific attainments, and as apparatus for making them is likely to be improved in the course of scientific research, it is not deemed advisable to include in this code any specific directions for making them.

RECORD OF THE TEST.

A "log" of the test should be kept on properly prepared blanks containing headings as follows :

TABLE NO. 58.

TIME.	PRESSURES.			TEMPERATURES.					FUEL.		FEED WATER.	
	Barometer.	Steam gauge.	Draft gauge.	External air.	Boiler room.	Flue.	Feed water.	Steam.	Time.	Pounds.	Time.	Lbs. or cu.ft.

REPORTING THE TRIAL.

The final results should be recorded upon a properly prepared blank, and should include as many of the following items as are adapted for the specific object for which the trial is made. The items marked with a * may be omitted for ordinary trials, but are desirable for comparison with similar data from other sources.

Results of the trials of a --
Boiler at --
To determine --

1. Date of trial -----------------------------------	
2. Duration of trial -----------------------------------	hours.
DIMENSIONS AND PROPORTIONS.	
(Leave space for complete description).	
3. Grate-surface ------wide------long------area ------	sq. ft.
4. Water-heating surface-----------------------------	sq. ft.
5. Superheating-surface -----------------------------	sq. ft.
6. Ratio of water-heating surface to grate-surface -----	
AVERAGE PRESSURES.	
7. Steam-pressure in boiler, by gauge-------------------	lbs.
*8. Absolute steam-pressure-----------------------------	lbs.
*9. Atmospheric pressure, per barometer---------------	in.
10. Force of draught in inches of water ---------------	in.
AVERAGE TEMPERATURES.	
*11. Of external air---------------------------------	deg.
*12. Of fire-room---------------------------------------	deg.
*13. Of steam---	deg.
14. Of escaping gases-----------------------------------	deg.
15. Of feed-water -----------------------------------	deg.
FUEL.	
16. Total amount of coal consumed--------------------	lbs.
17. Moisture in coal-----------------------------------	per cent.
18. Dry coal consumed --------------------------------	lbs.
19. Total refuse, dry------pounds=---------------------	per cent.
20. Total combustible (dry weight of coal, item 18, less refuse, item 19)-----------------------------------:----	lbs.
*21. Dry coal consumed per hour-----------------------	lbs.
*22. Combustible consumed per hour---------------------	lbs.
RESULTS OF CALORIMETRIC TESTS.	
23. Quality of steam, dry steam being taken as unity.	
24. Percentage of moisture in steam-------------------	per cent.
25. Number of degrees superheated--------------------	deg.
WATER.	
26. Total weight of water pumped into boiler and apparently evaporated-----------------------------------	lbs.
27. Water actually evaporated, corrected for quality of steam ---	lbs.
28. Equivalent water evaporated into dry steam from and at 212° F.---	lbs.
*29. Equivalent total heat derived from fuel in B. T. U.	B. T. U.
†30. Equivalent water evaporated into dry steam from and at 212° F. per hour-------------------------------	lbs.
ECONOMIC EVAPORATION.	
31. Water actually evaporated per pound of dry coal, from actual pressure and temperature-------------	lbs.
32. Equivalent water evaporated per pound of dry coal, from and at 212° F-----------------------------------	lbs.
33. Equivalent water evaporated per pound of combustible from and at 212° F---------------------------	lbs.
COMMERCIAL EVAPORATION.	
34. Equivalent water evaporated per pound of dry coal with one-sixth refuse, at 70 lbs. gauge pressure, from temperature of 100° F. = item 33×0.7249 pounds ---	lbs.

† Corrected for inequality of water level and of steam pressure at beginning and end of test.

RATE OF COMBUSTION.

35.	Dry coal actually burned per square foot of grate-surface per hour		lbs.
*36. *37. *38.	Consumption of dry coal per hour. Coal assumed with one-sixth refuse.	Per sq. ft. of grate surface	lbs.
		Per sq. ft. of water heating surface	lbs.
		Per sq. ft. of least area for draught.	lbs.

RATE OF EVAPORATION.

39.	Water evaporated from and at 212° F. per square foot of heating surface per hour		lbs.
*40. *41. *42.	Water evaporated per hour from temperature of 100° F. into steam of 70 lbs. gauge pressure.	Per sq. ft. of grate-surface	lbs.
		Per sq. ft. of heating surface	lbs.
		Per sq. ft. of least area for draught.	lbs.

COMMERCIAL HORSE-POWER.

43.	On basis of 30 lbs. of water per hour evaporated from temperature of 100° F. into steam of 70 lbs. gauge pressure (34½ lbs. from and at 212°)	H. P.
44.	Horse-power, builders' rating, at square feet per horse-power	H. P.
45.	Per cent. developed above, or below, rating	per cent.

NOTE.—Items 20, 22, 33, 34, 36, 37, 38 are of little practical value. For if the result proves to be less satisfactory than expected on the actual coal, it is easy for an expert fireman to decrease No. 20 by simply raking out some partly-consumed coal in cleaning fires, and thus making a fine showing on that simply ideal or theoretical unit, the "pound combustible." The question at issue is always what can be done with an *actual* coal, not the "*assumed* coal" of items 34, 36, 37 and 38.

E. D. M.

Hauling a 250 H. P. Heine Boiler up a Mountain.

Equitable Building,
DENVER, COLO.
Equipped with 750 H. P. Heine Safety Boilers.

CONDENSATION OF STEAM IN PIPES.

When steam pipes are exposed to the open air, the steam condenses more or less rapidly, according to the condition of the surfaces and the temperature and rate of motion of the air. This loss is quite serious in itself and further increases the losses by cylinder condensation, as indicated on page 77.

Experiments made by different parties in still air gave the following results:

TABLE NO. 60.

Condensation in Uncovered Pipes.

OBSERVER.	Difference of Temperature of Steam and Air.	Steam Condensed per Square Foot per Hour, per 1° F.	H. U. Lost per Square Foot per Hour, per 1° F.
Tregold	161° F.	0.0022 lb.	2.100
Burnat	196.6° F.	0.0030 lb.	2.864
Clément	151° F.	0.00217 lb.	2.071
Grouvelle	168° F.	0.0020 lb.	1.909
Average for steam of 20 lbs. absolute pressure	169° F.	0.00235 lb.	2.236

We further give an abstract of the results of a careful series of tests made by Mr. George M. Brill, M. E., in 1895, with the best modern coverings, and with the most accurate instruments. The steam pressure carried ran between 110 and 119 lbs. per square inch, and the temperature of the air varied from 50° to 81° F. in the various tests.

For the purposes of these tests about 60 feet of standard 8-inch wrought pipe, coupled together, in order to make it smooth and regular, was suspended where it could not be subjected to currents of air. In order to get the steam as dry as possible it was sent through a separator on its way to the test pipe, and in the short connection between the separator and the pipe was placed a throttling calorimeter. The test pipe had an inclination of one foot in its entire length, which insured drainage of all the water of condensation to the lower end, at which point the receiver was connected, and into which the water gravitated as rapidly as formed. The water was measured in this receiver, which consisted of four feet of 12-inch pipe, with graduated water glasses attached near the top and bottom. The same volume of water was allowed to collect each time, was measured under the steam pressure, and blown from the receiver at the end of the run. A careful determination was made of the amount of water collected by weighing the same volume while cold, and correcting for difference in weight due to the difference in temperature for the respective runs.

The tests were made upon a scale large enough—in fact, upon a pipe of the size and length which is very common in the average power plant—with sufficient care, and in a manner to insure accuracy in the results obtained, and are consequently of much interest and value to all users of steam.

The results reduced to the proper units are given in Table No. 61 below, and may be taken as fairly representative of the best modern practice. Of course, whenever steam pipes are placed where they are exposed to currents of air, the amount of condensation will be much greater than the tabular numbers.

This table also gives the saving in pounds of steam, and in dollars and cents due to the use of coverings. This saving is based on the assumption that coal costs $2.44 per ton, and adding 12 per cent for cost of firing, and taking 7 lbs. water per lb. of coal as an evaporative figure, which are rough approximations to average American conditions.

TABLE NO. 61.

Showing Radiation Due to Bare and Covered Pipes, and Saving Due to Coverings.

KINDS OF COVERING.	B. T. U. Transmitted per Hour per Square Foot Pipe per Degree Difference in Temperature.	Lbs. Steam Condensed per Hour per Square Foot Pipe per Degree Difference in Temperature.	Lbs. Steam Saved per 100 Square Feet Pipe per Year.	Saving in Dollars per 100 Square Feet Pipe per Year.
Bare Pipe	2.7059	.003107		
Magnesia	.3838	.000432	635,801	$110.82
Rock Wool	.2556	.000285	670,666	116.90
Mineral Wool	.2846	.000311	662,957	115.55
Fire Felt	.5023	.000591	603,389	105.17
Manville Sectional	.3496	.000409	645,174	112.45
Manville Sectional and Hair Felt	.2119	.000243	682,930	119.03
Manville Wool Cement	.3448	.000410	646,488	112.68
Champion Mineral Wool,	.3166	.000364	654,197	114.03
Hair Felt	.4220	.000472	625,376	109.00
Riley Cement	.9531	.001089	479,960	83.66
Fossil Meal	.8787	.001010	500,284	87.20

The presence of sulphur in the best coverings and its recognized injurious effects, makes it imperative that moisture must be kept from the coverings, for if present, will surely combine with the sulphur, thus making it active. This could be stated in other words, *keep the pipes and covering in good repair*. Much of the inefficiency of coverings is due to the lack of attention given them; they are often seen hanging loosely from the pipe which they are supposed to protect.

All coverings should be looked after at least once a year and given necessary repairs, refitted to the pipe, the spaces due to shrinkage taken up, for little can be expected from the best non-conductors if they are allowed to become saturated with water, or if air currents are permitted to circulate between them and the pipe.

As a very rough approximation we may say that each 10 square feet of uncovered pipe will condense, in winter, 105 lbs. of steam during a day of ten hours. Under the same conditions, the same pipe protected with the best covering will condense approximately 8½ lbs. steam.

In summer these figures will be reduced respectively to 80 lbs. and 6½ lbs. of steam.

Moisture in steam at the end of a long pipe line is often erroneously attributed to priming of the boiler; whereas, it is really due to condensation. The amount of steam condensed is really but a very small proportion of the total steam passing through the pipe, but gradually collecting at some point in the line, it is carried along in a body at intervals, producing the effects of entrained water.

Denver Consolidated Electric Light Co.,
DENVER, COLO.
Contains 3500 H. P. of Heine Boilers.

CHIMNEYS AND DRAFT.

According to Data and Rules given in our article on *Combustion* (p. 13, etc.), we find that from 12 to 14 lbs. of air are required per pound of coal. Anthracites require the least, bituminous coals more in proportion to their excess in volatile constituents. Most authorities consider a surplus of air requisite for complete combustion, so that a total amount varying from 18 to 24 lbs. of air per pound of coal is advised by various authors.

Taking 13 lbs. as the average amount of air chemically required, and the air at 62° F. and chimney gases at 500° F., this means that in order to attain perfect combustion we must sacrifice from 6 to 12 per cent of the calorific value of every pound of coal we burn in drawing " surplus " air through the furnace. Besides this, there is a loss in the cooling of the gases, and thus lessening the quantity of heat transmitted to the boiler. A thorough mixture of the air and the coal gas would do away with the necessity of most of this surplus air and thus prevent these losses. We have seen (pp. 14, 15) that an increase in the rate and temperature of combustion reduces the proportion of surplus air required. This means reduced grate area and increased draft, and points to high chimneys.

What we call draft is simply the fall of the heavier (because colder) outside air to supply the place of the lighter (because heated) gases which rise from the furnace to escape through the chimney. We cause it artificially in a furnace just as wind is caused by the heat of the sun in nature.

The difference in weight of the column of hot gas in the chimney and that of a column of the outside air of the same height is the force which causes the draft.

It is customary to measure the draft in inches of water. We will assume the external air to be at 62° F. and that in the chimney at 500° F. A cubic foot of air at 62° F. weighs 0.0761 lbs.; and at 500° it weighs 0.0413 lbs.; the difference is 0.0348 lbs. For a chimney 100 ft. high we would have on every square foot of its cross section at the bottom an upward pressure of 100 times 0.0348 lbs. = 3.48 lbs. A cubic foot of water at 62° F. weighs 62.32 lb., i. e., a column of water 12″ high exerts a pressure of 62.32 lbs. per square foot on its base; 1″ of water therefore means a pressure of 5.193 lbs. on a square foot or one of 0.577 ounces on a square inch. Our 100-ft. stack therefore shows a draft of 3.48 ÷ 5.193, equals 0.67 inches of water or about 0.39 ounces of pressure per square inch.

In the above we have considered the gases in the stack as of the same specific gravity as air. But this is not true. The chimney gases are a mixture of carbonic acid gas nitrogen and gaseous steam, complete combustion being assumed.

Carbonic acid gas has a specific gravity of 1.529; nitrogen of 0.971; steam of 0.624; air being taken as the basis = 1.

Hence in place of air at 500° F. weighing 0.413 lbs. per cubic foot, we have a mixture of gases whose weight varies with the varying amounts of

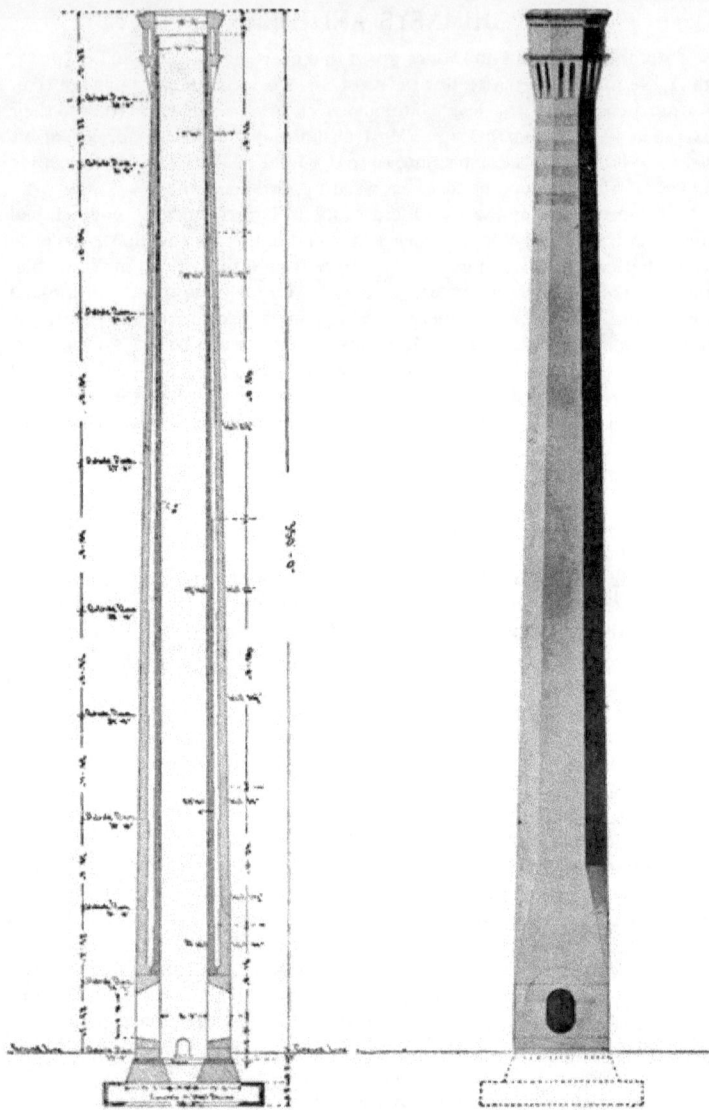

Brick Chimney at the Omaha & Grant Smelting and Refining Works,
DENVER, COLO.
Designed by Wm. M. Scanlan.

each constituent. These differ with different coals, and therefore different kinds of coal will cause differences in the draft of a given chimney, even when the temperatures involved are the same. The following table gives for five well known coals the number of pounds of air required per 100 lbs. of coal burnt, weights of the resultant gases, the number of cubic feet of chimney gases at 500° F. and the weight per cubic foot of the mixture, at this temperature in the chimney.

TABLE NO. 63.

KIND OF COAL	Reference Letter.	Per Cent.				Per 100 lbs. Coal.			
		Moisture.	Fixed Carbon.	Volatile Matter.	Ash.	Air necessary for complete combustion.	Total weight of chimney gases, lbs.	Cubic feet chimney gases at 500° F.	Weight per cubic foot chimney gases at 500° F.
Anthracite (Pa.).......	A	1.81	86.75	6.18	5.26	1279	1374	31440	0.0437
New River (Bit.)......	NR	....	77.00	18.00	5.00	1385	1480	34454	0.0429
Youghiogheny "	Y	2.00	59.00	33.00	6.00	1448	1542	36367	0.0424
Mt. Olive "	MO	6.80	46.00	37.00	10.20	1353	1443	34711	0.0416
Collinsville "	C	9.00	32.00	46.00	13.00	1345	1432	35052	0.0408

These different weights of the gases of combustion then cause differences in draft power of the same chimney, even when the temperatures of the gases and of the outer air are the same in all cases. Table No. 64 is figured for certain average conditions of practice. The last line is added to show the results as usually figured on the assumption that the chimney gases have the same weight as air.

TABLE NO. 64.

Draft Pressures Due to Different Coals, with Different Temperatures of Air, but same Chimney Temperature. Chimney 100 Feet high above Grates.

Gases of Combustion from	Weight 1 Cubic Foot at 500° F.	Weight 1 Cubic Foot Air at 0° F.	Draft in inches of Water.	Weight 1 Cubic Foot Air at 62° F.	Draft in inches of Water.	Weight 1 Cubic Foot Air at 102° F.	Draft in inches of Water.
A	0.0437	0.0864	0.822	0.0761	0.624	0.0707	0.520
N. R......	0.0429	0.0864	0.837	0.0761	0.639	0.0707	0.535
Y	0.0424	0.0864	0.847	0.0761	0.649	0.0707	0.545
M. O.....	0.0416	0.0864	0.863	0.0761	0.664	0.0707	0.560
C	0.0408	0.0864	0.878	0.0761	0.680	0.0707	0.576
Air	0.0413	0.0864	0.869	0.0761	0.670	0.0707	0.565

The table further shows that difference in temperature of the outer air may affect the draft of the chimney to the amount of 50 per cent and over. In practice we find sometimes too little air, which shows inexcusably bad design or management, sometimes (though rarely) just enough, and some-times (see p. 15) amounts of surplus air varying from 10 per cent. to 100 per cent. In the former case we have imperfect combustion which may mean a waste of the entire volatile portion of the fuel, which by Table 63 may run up to 20 per cent and more of actual loss.

In the other cases we have to draw into the furnace, heat and expel through the chimney varying quantities of inert air, which again represent various percentages of loss. The following table illustrates this :

TABLE NO. 65.

Showing Weight and Volume of Chimney Gases from 100 lbs. each of Various Coals at 500° F. on the Assumption of Various Percentages of Surplus Air.

Ref. Letter	10% SURPLUS AIR.			25% SURPLUS AIR.			50% SURPLUS AIR			100% SURPLUS AIR.		
	Wt.	Wt. per Cub. Ft.	Vol. Cub. Ft.	Wt.	Wt. per Cub. Ft.	Vol. Cub. Ft.	Wt.	Wt. per Cub. Ft.	Vol. Cub. Ft.	Wt.	Wt. per Cub. Ft.	Vol. Cub. Ft.
A...	1502	0.0435	34540	1694	0.0434	39190	2014	0.0429	46940	2653	0.0425	62440
N.R.	1619	0.0428	37814	1826	0.0427	42808	2172	0.0418	51154	2865	0.0416	67854
Y...	1687	0.0420	40187	1904	0.0419	45447	2266	0.0418	54207	2990	0.0417	71747
M.O.	1578	0.0416	37981	1781	0.0415	42891	2119	0.0415	51071	2796	0.0414	67431
C...	1567	0.0409	38322	1768	0.0409	43182	2104	0.0410	51310	2777	0.0411	67570

If we take for example Youghiogheny coal, we see that with 100 per cent surplus air the weight of the chimney gases has been reduced to 0.0417 lbs. per cu. ft. We have, then with the air at 62° F., a draft pressure of 0.66 inches in place of the 0.649 inches of Table 64. That is a gain of 1½ per cent in draft by admitting 100 per cent surplus air; but we have 96 per cent more in volume of gases to push through the chimney. If we still assume the temperature of chimney gases at 500° F., this surplus air (at 0.2379 specific heat) requires 150592 H. U. to bring it from 62° to 500°. As this Youghiogheny coal averages 12800 H. U. per lb., it would take all the heat from 11.76 lbs. of coal to heat this surplus air, a loss of nearly 12 per cent in the efficiency or economy.

If on the other hand we assume that the chimney temperature will be reduced, and no fuel is wasted in heating this surplus air, this total possi-ble reduction based on the same at 62° F., with the specific heat of air at 0.2379, that of the gases of combustion at 0.2495, and that of the mixture at 0.244, amounts to 207° F., entailing practically the same loss in heat, viz., 151100 H. U. But with the chimney temperature at only 293° F. we would have only 0.023 difference in weight of inside and outside columns, cr 0.44 inch draft, in place of 0.65 inch, a loss of over 32 per cent in chimney efficiency or capacity. In other words this surplus air has reduced the velocity of the gases in the chimney nearly one-third, while giving us 96 per cent more gases to move. This shows forcibly that a low chimney temper-

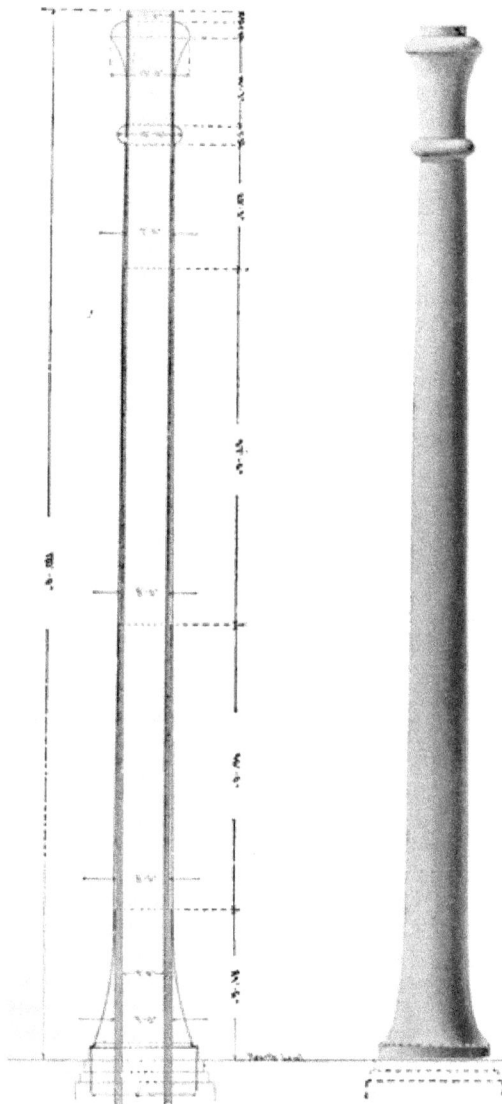

Example of Iron Chimney.
Designed by J. P. Withrow.

ature *may* show waste of fuel; it shows economy only when attained with a minimum of surplus air.

The velocity per second of the gases in the stack is given by the formula $V = \sqrt{2gh}$ in which "h" is the height of a column of the hot chimney gas whose weight is equal to the difference in weight of the air outside and the gas inside of the chimney. As we can express this head in inches of water "p," we get the formula $V = C\sqrt{p}$ in which the constant "C" varies according to the composition of this gas. For the gases at 500° F. from the various coals above considered, the formula becomes:

$$V = 87.2 \ \sqrt{p} \quad \text{for Anthracite Coal.}$$
$$V = 87.92 \ \sqrt{p} \quad \text{for New River Coal.}$$
$$V = 88.56 \ \sqrt{p} \quad \text{for Youghiogheny Coal.}$$
$$V = 89.36 \ \sqrt{p} \quad \text{for Mt. Olive Coal.}$$
$$V = 90.24 \ \sqrt{p} \quad \text{for Collinsville Coal.}$$

For the entrance velocity of the air under the grate, we have for 62° F. the formula $V = 66.1 \sqrt{p}$

These formulas give us velocities of 75 ft. p. second and over for the quite ¡usual draft pressure of 0.75 inch of water. But no such velocities exist in boiler chimneys. The reason is that only a small part of that difference in pressure, which our draft gauge measures at the base of the stack is or can be utilized for producing velocity. The greater part of it is required to overcome the frictions of the grate with its bed of fuel, and that of the boiler flues or tubes. The ignoring of this fact has led to the oft repeated error that there is practically no gain in chimney capacity by an increase in the temperature of the gases, because their increase in volume counterbalances the increment in velocity. And thus the *maximum capacity* is stated as reached when the gases have about double the volume of the external air. On the other hand, an English authority, Mr. Thos. Box, shows that with a flue 100 ft. long from furnace to base of chimney, the maximum power or capacity is reached only when the gases in the stack have about 3½ times the volume of the external air, i. e., when their temperature has risen to nearly 1400° F. Neither of these views recognize that the character of the fuel, the thickness of the bed upon the grate, the methods of firing, and the proportions of the grate are really the determining factors in this question. And while it is true that temperatures as high as 1100° F. have been observed in practice, they show very *bad practice*. But even in much more moderate limits an increase of stack temperature may materially increase the power or capacity of a given stack.

Careful experiments are sadly needed for determining what fractional parts of the draft are expended in overcoming the various frictions mentioned. But from a large number of boiler tests we may safely figure out that modern practice requires entering velocities of from 9 to 25 ft. per second for the air, and escaping velocities of from 7 to 30 ft. for the chimney gases; and with due allowance for chimney frictions, we have then a total of from 0.03 to 0.22 inch of draft required for these. The frictions in furnace and boiler are similarly found to run from 0.4 to 0.6 inch, making the totals range from 0.43 to 0.82 inch. With these data in hand we can figure out the probable effect of high chimney temperature in increasing the actual working power of a stack.

We will assume a plant with a chimney 100 ft. high, burning Youghiogheny coal at a pretty brisk rate, taking 50 per cent surplus air, and chimney gases at 500° F. and air at 62° F. The stack at this rate is doing its duty well, and the plant is fairly economical. A demand for one-third more steam is made by those little additions to the machinery or increased direct use of live steam, which in the popular belief "cost nothing when you once have a good boiler." *The boiler and the fireman have to get this steam some how.* The only recourse will be such changes in the method of firing as will burn more coal per minute, and the only way to do it is by letting the gases escape hotter and thus get the increased draft. By firing oftener and more judiciously, the bed of fuel will not be much thickened and the friction here will be increased probably only one-fourth, and in the flues hardly that much.

Suppose the chimney gases to go to 900° F. then the account will stand about as follows :

	Ordinary Work.	Work Increased.
In percent	100%	133⅓%
Stack temperature	500°	900°
Air	62°	62°
Available draft	0.649 inch.	0.889 inch.
Air entering at velocity of	10 ft. p. sec.	13.3 ft. p. sec.
Gases escaping at velocity of	12 ft. p. sec.	23.0 ft. p. sec.
Draft required for entering velocity	0.0230 inch.	0.0400 inch.
Draft required for escaping velocity	0.0182 "	0.0676 "
Draft required to overcome furnace frictions	0.6000 "	0.7500 "
Total expended	0.6412 "	0.8576 "
Leaving balance available	0.0078 "	0.0314 "
Total pounds gas from 100 lbs. and 133 lbs.coal with 50% surplus air	2266 lbs.	3021 lbs.
Total volume at 500° and 900°	54207 cu. ft.	101376 cu. ft.

As these volumes bear to each other the same ratio as the velocities 12:23, the stack is now doing its work just as well as before. In fact the balance of draft remaining could be used in increasing the velocity of exit to nearly 28 ft., i. e., carrying off nearly 22 per cent more gas in volume, equivalent to a further increase in capacity for coal burning of nearly 16 per cent. Or practically we *can* increase the capacity or power of the stack by nearly fifty per cent by increasing the temperature of the gases from 500° to 900° F. The *cost* of doing this is of course *very great.*

At 500° the chimney required for its total work of drawing in the air and expelling the gases about 13 per cent of the fuel burnt; at 900° it requires 25 per cent, a clear loss or *waste* of 12 per cent.

The same result can be attained without a pound of additional fuel by raising the chimney 40 ft.

Table No. 66 illustrates this general question, but in applying it to any existing problem, careful measurements should first be made of existing resistances on the way from boiler front to base of chimney.

Showing Changes in Capacity of Chimney by Changes in Temperature of Gases, With Height Constant; or Changes in Height with Temperature Constant. Air at 62° F. Weight of Gases, the Average of the Five Coals Considered.

	400°	500°	600°	700°	800°	900°
Temperature of escaping gases with 100 ft. chimney	400°	500°	600°	700°	800°	900°
Per cent of total coal necessary to establish draft	10	13	16	19	22	25
Draft obtained in inches of water	0.56	0.65	0.73	0.79	0.85	0.89
Height of chimney for same draft, at 500° F., in feet	86	100	112	121	131	137

We append a further table showing the effect on velocities and areas of chimneys from differences in quantities and mixtures of gases, and from the varying values as boiler fuels of the five coals considered. While this is figured on the basis of no surplus air, the ratios found will be but little affected by such surplus.

TABLE No. 67.

	Anthracite.	New River.	Yonghiogheny.	Mt. Olive.	Collinsville.
Velocities in % of A	100	101	102	103	104
Quantities of gas in cub. ft	31440	34454	36367	34711	35052
Areas should be, in % of A for equal quantities of coal	100	108.5	113.4	107.2	107.2
Comparative evaporative efficiency in lbs water from and at 212°	9	10.5	10	7.5	7.
Pounds coal burnt to be equal in effect to 100 pounds A	100	85.7	90	120	128.5
Equivalent chimney areas %	100	93	102	128	138

The above considerations show the practical difficulties in the way of any general formulas for chimney height and area, and explain why the "doctors disagree" in regard to them. If we had exhaustive and complete tests on the amount of grate and fuel bed frictions under the severe conditions of modern boiler practice, and with different kinds, qualities and conditions of coal, probably all accepted formulas would, by substitution of new constants, be brought into substantial accord. But constants based on grates with 25 to 33 per cent air space, and on a consumption of 8 to 15 lbs. coal per hour per square foot of grate will lead to erroneous results in modern practice with 50 per cent air space and a consumption of 20 to 40 lbs. coal. Therefore our results must be modified by careful judgment based on well known local conditions. The best known formulas are Smith's, Kent's and Gale's. They are as follows:

Smith.

$$A = \frac{0.0825\,F}{\sqrt{h}}$$

$$h = \left(\frac{0.0825\,F}{A}\right)^2$$

Kent.

$$A = \frac{0.06\,F}{\sqrt{h}}$$

$$h = \left(\frac{0.06\,F}{A}\right)^2$$

Gale.

$$A = 0.07\,F^{\frac{3}{4}}$$

$$h = \frac{180}{t}\left(\frac{F}{G}\right)^2$$

In which "A" = area, "h" = height of stack in feet, "F" = pounds

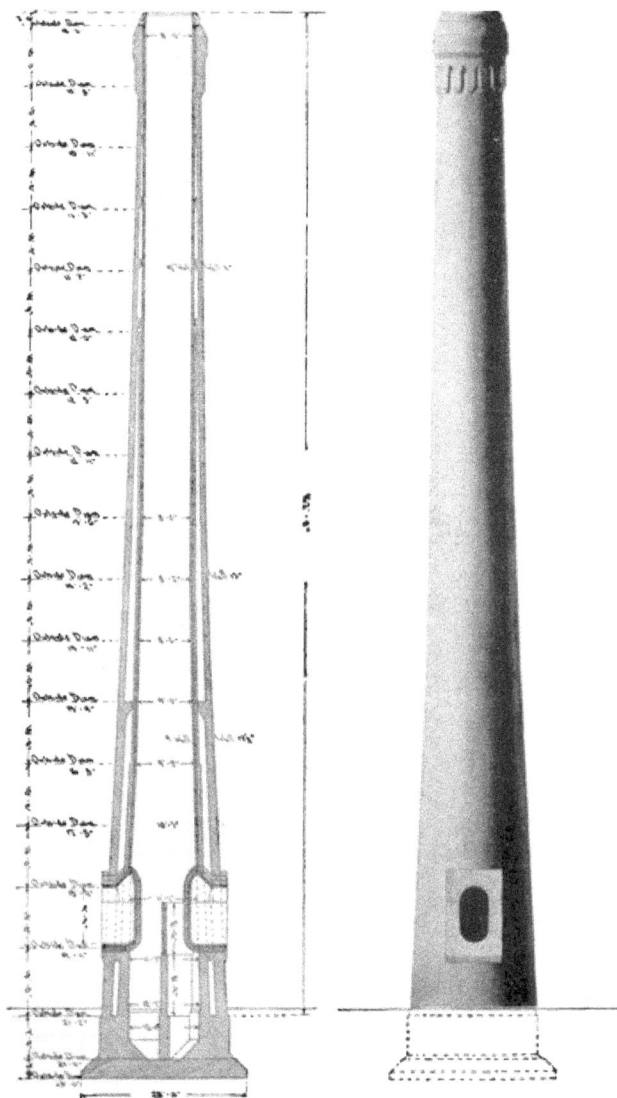

Brick Chimney at the Power House of the Union Depot Ry. Co.,
ST. LOUIS, MO.
Designed by E. D. Meier, M. E.

coal burnt per hour, "t" = the stack temperature, and "G" = grate area. But in Kent's formula, "A" represents the effective area only, and he adds a ring 2" wide all around to allow for chimney frictions. Thus if the formula gives you a chimney of 41" diameter or of 36" square, you must make its actual size 45" diam. or 40" square. For 100 ft. height, Kent's formula gives a total area 11 per cent larger than Smith's for 250 lbs. coal per hour (50 H. P.); exactly the same for 500 lbs. coal (100 H. P.); 18 per cent smaller for 1000 lbs. (200 H. P.); 24 per cent smaller for 5000 lbs. (1000 H. P.) etc. The 5 lbs. coal per H. P. is merely a convenient assumption, and is based on an evaporation of 7 lbs. water per lb. of coal. The areas will vary according to the quality of coal, and such data on evaporation as local practice supplies, as indicated by our Table No. 67.

Kent's formula has the advantage of recognizing the practical fact that for larger powers the area of chimney required per horse power becomes less.

The general form of Gale's formulas is more promising. But as his constants are based on observed data much smaller than those of best modern practice, they lead to rather too large results. But his making the height depend only on the stack temperature and the rate of combustion is much more in accord with the facts than making height and area interdependent as the other two formulas do. With Gale's constants modified so that $h = \frac{120}{t} \left(\frac{F}{G} \right)^2$ the heights can be fixed and then Kent's formula for areas applied. The interdependence of height and area exists only in limits defined by practical observation. Outside of these the assumption leads to an absurdity. F. i. Kent's formula for area would give a 64" chimney 9 ft. high as equivalent to a 35" chimney 100 ft. high.

Practical and local considerations generally fix the height required. The chimney must be higher than surrounding buildings or hills, else whenever the wind comes from the direction of the higher object, the draft will be seriously impaired. Then the nature of the coal must be considered.

Mr. J. J. de Kinder, M. E., who has been engaged on a large number of boiler and coal tests for the Pa. R. R. and other large consumers, using telescopic stacks to meet this very question, gives 75 ft. as height for the most free-burning bituminous coals, 115 ft. for slow-burning bituminous, and from 125 to 150 ft. for anthracite coals. These latter being of three kinds, free-burning such as Lykens Valley; semi-free-burning such as Delaware and Lackawanna; and hard-burning such as Lehigh Valley; they cannot be distinguished from each other by appearance.

DeKinder gives as necessary draft for anthracite 0.75 inch to 0.88 inch, and is in substantial agreement with Dr. Emery and Mr. Hague in this. He gives 20 to 25 lbs. per hour as minimum rates of combustion, 40 per cent air space in grates for anthracite and 50 per cent for bituminous coals.

We give in Table No. 68 appropriate heights and areas of chimneys for powers from 75 to 3100 horse-power; based on an assumed evaporation of 7 lbs. water per lb. coal, equivalent to 5 lbs. coal per H. P. per hour.

For better or poorer coals any figures from this table can be readily modified by referring to the tables in the earlier pages of this article.

If bituminous slack is to be used, the chimney should not be less than

100 feet high, and not less than 125 feet high for anthracite pea, or 150 feet for anthracite buckwheat.

TABLE NO. 68.

Area Square Feet.	Diameter, Inches.	HEIGHTS IN FEET.												
		75	80	85	90	95	100	110	120	130	140	150	175	200
		COMMERCIAL HORSE POWER.												
3.14	24	75	78	81	....	....	....	....	....	....	....	....	....	....
3.69	26	90	92	95	98	....	....	....	....	....	....	....	....	....
4.28	28	....	106	110	114	117	120	..	....	....	....	....	....	....
4.91	30	....	122	127	130	133	137	....	....	....	....	....	....	....
5.59	32			144	149	152	156	164	....	....	....	....	....	....
6.31	34			162	168	171	176	185	....	....	....	....	....	....
7.07	36				188	192	198	208	215	....	....	....	....	....
8.73	40					237	244	257	267	279	....	....	....	....
10.56	44					287	296	310	322	337	....	....	....	....
12.57	48						352	370	384	400	413	....	..	....
15.90	54						445	468	484	507	526	....	....	....
19.63	60							577	600	627	650	672	....	....
23.76	66							697	725	758	784	815	....	....
28.27	72								862	902	932	969	1044	....
38.48	84								1173	1229	1270	1319	1422	....
50.27	96									1584	1660	1725	1859	1983
63.62	108									2058	2102	2181	2352	2511
78.54	120										2596	2693	2904	3100

Whenever it becomes necessary to have long flues leading to a chimney, the power of the latter becomes more or less impaired. We adapt the following table from Mr. Thos. Box; the *total* length of flue from grate to base of chimney must be considered.

TABLE NO. 69.

Reduction of Chimney Draft by Long Flues.

Total length of flues in feet...	50	100	200	400	600	800	1000	2000
Chimney draft in percent	100	93	79	68	58	52	48	35

A further loss in draft results from any downward course of the gases in the flue. It may be roughly accounted for by using double the length of such down turn in making up the total flue lengths for the above table.

Where several boilers lead into one chimney, a further factor comes in to reduce the required area. The heaviest work for the chimney is just after firing, since the friction through the fresh coal is greater and the temperature less than some minutes later. But it would be very bad practice to fire all boilers or all doors simultaneously. Hence the second and succeeding boilers do not require as much area as the first. It will be safe to figure 75 per cent for the second and 50 per cent each for the third, fourth, etc. But it is advisable to increase the height slightly for each boiler added.

E. D. M.

1200 H. P. Plant of Heine Boilers at Central Distillery,
ST. LOUIS, MO.

CONCENTRATION AND DISTRIBUTION OF POWER.

From the time that man first began to call the forces of nature to aid him in his handicraft, there has been a gradual increase in the size of power plants. At each stage of progress it became plainer that *power, repairs and labor could be saved* by larger wind or water wheels, turbines, and finally steam engines. The best engineer or millwright costs less than two of somewhat less ability, duplicate parts for one machine serve for prompt repairs as well for one out of twenty as for one out of three, a 500-horse power engine costs less than two of 250-horse power, etc., etc.

But there were other causes which imposed limits which it was disastrous to pass. The laborers grew in number in some ratio with the increase in power; they must live near the works. Often the *best place* for the power plant was the *worst place* for them. The conversion from *heat energy* to *mechanical force* frequently demands a site at low water level difficult of access or unhealthy for the laborers. *Shafting* and even *wire cables* have short distance limits for the economical transmission of power.

But with the development of *electrical power*, which commenced in the last decade, and is advancing now in almost geometrical ratio, very large steam plants have multiplied. It becomes possible to develop an *immense amount of power* in one place, since with but one more conversion—from mechanical into electrical energy—we can send it, divided into just such quantities as fit each time and place, *to points many miles apart*, with losses exactly controllable. Locations can therefore be chosen *where fuel and water are cheapest*, where the refuse is easiest disposed of, etc., and where every item of *economy*, multiplied by the enormous quantities involved, becomes a question of grave concern and careful calculation. In these large and essentially *modern power plants* will be found, as peer of the best type of the compound condensing or triple expansion engine, the *Modern Water Tube Boiler*. When in a plant like these, the old fire tube type of boiler is found, it is but an *exception which proves the rule*. As an apt illustration of this development we may compare the *Centennial Exposition at Philadelphia* in 1876 where a single 1000-horse power engine sufficed to drive all the machinery, and where fire tube boilers were the rule and water tube the exception, with the great *Columbian World's Fair* of 1893 at *Chicago*, where an installation of *15,000-horse power boilers* becomes necessary, from which condensing engines of the compound and triple expansion type will develop about 25,000-horse power, two-thirds of which will be converted into electrical energy.* All these boilers are of the *Modern Water Tube type, space, safety, economy and æsthetic considerations* having barred the others.

The detailed and painstaking investigation into all the points involved in steam making, which in such large plants precedes and influences the design, and the choice and size of the parts, is of course not possible in small plants, whose owners must necessarily follow on the lines marked out by these large and successful installations. But they will the more readily profit by such experience when prepared to analyze the different elements which together compose a modern boiler plant. As a guide to such analysis we offer in the following pages a few elementary thoughts on the salient points involved.

*By general consent a horse power in a boiler is considered as the evaporation into steam, at seventy pounds gage pressure, of thirty pounds of water per hour, as being about the quantity a good slide valve engine requires. A good single cylinder Corliss engine uses only twenty-five pounds, a compound condensing eighteen to twenty pounds, and a triple expansion thirteen to sixteen pounds. This explains the apparent discrepancy between boiler and engine power.

ANHEUSER-BUSCH BREWING ASS'N.

Equipped with 5200 H.P. Alsen Safety Boilers

A MODERN BOILER PLANT.

A good boiler plant is something essentially modern. Since *Watt* yoked the *Power*, and *Stephenson* harnessed the *Speed* of *Steam* to the triumphal car of modern progress, invention has been busy, throughout the civilized world, with improvements in all the elements of a complete steam plant.

But owing partly to the fact that the engine seemed to offer more chances for experiment, and better opportunity for observation, and partly to the knowledge that the losses in the engine were vastly greater than in even a carelessly designed boiler plant, the engine has received by far greater attention. Even now it is not an unusual thing to find a steam plant in which *every refinement of modern engineering* has been carefully brought to bear in the design and construction of engine and shafting, while the boiler plant has been settled by prescribing the number of square feet of heating surface, and adding a few commonplace specifications about the steel, which can be as well filled by a high sulphur steel as by good flange stock. Many an intelligent manufacturer will point with pride to his polished Corliss engine, will show you model indicator cards from it, while neither he nor his engineer can tell you within 25 per cent what his boilers are doing.

It is not uncommon to find the boilers stowed away in some hole, so close, dark and ill-ventilated that no self-respecting skilled laborer *will continue to work in it*, and a good fireman is emphatically a skilled workman, having charge of an important chemical process whose proper handling, in many lines of manufacture, determines whether the books will show loss or profit at the end of the year.

Naturally enough, ill-designed, badly proportioned breechings or flues are often found in such places, connecting into chimneys neither wide enough nor high enough for the work expected of them. But within the last decade *more attention has been given to the boiler plant*. Much educational work has been done by boiler companies, notably by one which annually publishes in its catalogue much useful information and many convenient tables of data connected with steam generation, which are not elsewhere readily available to the average steam user or his engineer. Much credit is due to the large electrical companies who have boldly departed from *antique superstitions*, and have put as much thought into their boiler plants as into the other elements of their large installations.

A boiler plant consists in the main of *three essential parts*, each one of which has its own important office in the success of the whole.

First, there is the *Chimney* or *Stack* with its *Flue* or *Breeching*, to carry off the waste gases and to create the *Draft*, without which combustion in a practical and economic sense is impossible.

Second, the *Furnace* or *Setting*, whose arrangement and dimensions determine the important elements of quantity and economy of *combustion*.

Third, the *Boiler*, whose proportions and design must be such as enable it to *absorb* the maximum amount of the heat produced by the furnace, thus determining finally the capacity and economy of the whole plant. These *separate and distinct offices* of the three component parts of a boiler plant are *often confounded*, not only by those to whom a boiler-room is simply a vague counterpart of the Black Hole of Calcutta, but even by those who claim to "know all about boilers." How often is the boiler manufacturer met by the question: "Will your boiler burn slack?" or "tanbark" or some other fuel desirable because cheap. Aside from the fact that the boiler has usually very little to do with it, the question can only be answered by exercising the Yankee privilege of asking a few more. F. i. "How much draft have you?" or "What are the dimensions of your chimney?" the answer will generally be "a splendid draft," or "we have a fine big chimney built only a few years ago." But this gives the boiler man but a very vague idea. *He wants facts* and he does not get them. The splendid draft may prove to be, according to the personal equation of his informant, anything from four-tenths of an inch to an inch of pressure, the chimney may be anything from half to full capacity for the work in hand, and yet upon an accurate knowledge of these data the correct answer to the first question depends.

THE CHIMNEY.

The Chimney determines how many pounds of fuel can be burnt per hour, the quantity varying with the kind of fuel in very narrow limits, and also to some extent depending on atmospheric conditions. Its office is to remove the waste gases whose quantity varies but little whether smoke accompanies combustion or not, and to supply enough air to oxydize all the fuel. The *Draft* pressure is simply the difference in weight between a column of hot and therefore light gas in the chimney, and a column of air outside, of the same height and area. The greater the draft pressure, the greater the speed of the spent gas leaving and the fresh air entering the furnace, and hence the greater the *quantity of fuel* which the same chimney area will enable us to burn.

This pressure, as explained, depends on the *height and temperature* of the column of waste gas; it may be increased at will either by making the chimney higher or allowing the spent gas to escape at a higher temperature. The latter method is very wasteful and should never be resorted to except where the former cannot, for some local reasons, be adopted. Of course, with larger chimney area less speed will suffice for the same quantities of gas and air, and this fact is often urged to bolster up the antique superstition that a *low chimney* with *ample area* will do the same work as a *tall one of less diameter*. If this were true, removing the roof of the boiler house ought to prove a good substitute for an expensive chimney, and a gas globe might conveniently replace the broken chimney of a student lamp.

It is just here that the nature of the fuel affects the matter. To cause combustion the air must be brought into *intimate contact* with all the particles of the fuel. With gas or oil this may be done with small initial draft. The *frictional resistance* to the passage of the air through a bed of solid fuel of any kind increases with the decrease in the size of the pieces, lumps or grain of the fuel. Hence a *sharper draft* is required for sawdust or tanbark than for cordwood, for slack or pea coal than for nut or egg coal. But the smaller the grain of the fuel the more surface is presented for the oxydizing action of the air, hence the more uniform the combustion. Therefore the careful fireman breaks his lump coal just before firing.

Again most coals have two rates of combustion which give *best economic results*. One usually a very low one and hence hardly available in the very limited space generally fixed by modern conditions. The other is a much higher one, the intermediate rates being frequently very wasteful. This *higher rate makes more power possible* in the minimum of floor area and hence meets modern demands. It developes higher temperatures, and, as great differences in heat favor its transmission, it makes more work possible in the boiler.

Finally a strong draft in the chimney is *less liable to interruption* by gusts of wind than a sluggish one. All these considerations point to the tall chimney as the source and fountain of all the energies of a modern steam plant.

The smoke stacks of the Pacific Mills, Lawrence; the Boston Edison Co.; the Narragansett Electric Light Co., Providence; Broadway Cable R. R. New York; Clark Thread Mills, Newark; Union Depot R. R., St. Louis; Chicago Edison Co., and Anheuser-Busch Brewery, St. Louis, are good examples of modern practice in the matter of tall chimneys.

The forty to sixty feet smoke stacks which were "plenty high enough" *belong to the past*, with the old stone mills, the ram shackle engines with the gothic ornaments, low steam and timber bed frames.

The Flue or *Breeching* connecting the furnace or setting to the chimney properly forms part of it. It should be of equal or slightly larger area and where changes in shape or direction cannot be avoided they must be made easy and gradual, carefully preserving the area at all points. *Abrupt turns or contractions* of area are known to *interfere* with the flow of liquids; frequent and facile observation shows this to every one, and tables are published showing the observed loss in effect by those of most common occurrence. In the case of gases the effect is even more damaging, since the initial force is generally (in a chimney *always*) limited, while opportunities for observing this action are not frequent and have to be specially created. Therefore so many sharp turns and sudden changes in area are met with in steam pipes and smoke flues, which, a little thought would prove, should be avoided. Where one chimney serves several boilers, the *branch* of the breeching or flue for each must be *somewhat larger* than its proportionate part of the area of the main flue.

Forced draft is sometimes employed with good success. It should be an *adjunct* merely, but cannot be made to replace a tall chimney. Combustion will not be as perfect under pressure as under a slight vacuum. A leakage of air inward through the furnace walls helps to supply hot air for combustion, and to some extent reduces and counteracts losses by radiation. But excessive forced blast which more than counterbalances the draft of the chimney will increase radiation and by leakage through the walls, doors, etc., outwards cause much loss. Worst of all it *interferes with the fireman* by making his work hard and unsatisfactory.

THE FURNACE.

The chimney having fixed the quantity of fuel we can burn, we must arrange our furnace so that it will do the best work within this limit. We must remember that *the draft must be husbanded*, its whole force to be called on only for our maximum effort. The *kind of fuel* and the nature of the service will determine the *proportions of our furnace*. The furnace which will give excellent results on coal will be found inadequate for wood, if it be proportioned for the steady and regular work of a flour mill, it must be modified to meet the sudden and varying demands of an electric railway. The *grate must*, in area, in width and shape of air spaces, in length and design of bars *be adjusted* to the kind of work the plant is to do, and the peculiarities of the fuel. Thus a baking and clinkering coal requires few and wide air spaces, a dry and friable one must have many and narrow ones. The total *air space* of the grate must be made as large as possible since it is the *active element;* the metal must be reduced in width as much as is compatible with strength. The surface of the grate must be as smooth and even as possible so as to offer no impediment to the use of the clinker bar and other fire tools. The longer time required for the perfect combustion of a fuel the larger must furnace, combustion chamber and flue be arranged. For sufficient air, high temperature, and time and space are equally important conditions of thorough combustion, and this must be completed before the gases are brought in contact with the heating (or here cooling) surfaces of the boiler. These rules apply to the various patent grates, stokers and furnaces as well as to the standard devices of established practice. And the best invention must in its application be supplemented by experience, calculation and design. The *walls of a good furnace* should have as *few openings*, doors, etc., as possible, since every break in the bond of the brickwork increases the tendency to cracks, which can never be entirely avoided, but which cause leaks so detrimental to complete economy. Double walls with air spaces between them should always be employed where practicable, so that this unavoidable indraft through the cracks may be heated and utilized for secondary combustion.

The lining of the furnace proper and the bridge wall should be made of a quality of fire brick which combines *great refractory power* with *hardness and toughness* to resist the abrasion due to the fire tools and the clinkers. The combustion chamber and flues may be lined with a cheaper grade since the heat is less and no abrasion possible. The cheap plan of using no fire brick abaft the bridge wall is wasteful in the end and therefore bad practice. As *no bond* of either fireclay or mortar is *absolutely reliable* under furnace temperature, long and stout anchor rods should be used to tie the walls securely together. It is of course necessary to make the joints between the furnace and the boiler as *nearly air-tight as possible.* This is best done by leaving joints wide enough to clear all projecting parts of the boiler, such as rivet heads, etc., and then filling them with some spongy material, f. i., tow or waste thoroughly saturated with fireclay. This is pliable enough to *follow the movements* caused by alternate expansion and contraction *without racking the brickwork* or impairing the joints. By this arrangement the boiler can be made entirely independent of the stability of the walls. For all clinkering coals a cemented ashpit kept full of water is advisable.

Having now designed a furnace, capable of burning our fuel to best advantage, little and slowly when the demand for power is slight, much and fiercely when the full load is put on, i. e., having devised the best means for waking the sleeping force in the fuel to the active energy of living *Heat*, we want means to translate this into *Mechanical Power.*

The *Steam Boiler* furnishes the means. If we except certain dangerous vapors, *steam*, which is the gaseous form of water, is the substance whose *expansive force* grows most rapidly with each increment of heat. It has therefore become to civilized man the almost universal means of drawing active working force from the latent Sun-Energy stored up for him for ages

1750 H. P. Heine Plant of Boilers in Union Depot R. R. Electric Plant,
ST. LOUIS, MO.

past by provident Nature. In the *furnace the energy of heat* has been called to life; the *boiler* is now to *absorb* this heat and to *transmit* it to the water within. This will first rise in temperature with less than five per cent expansion, until a point is reached when each additional unit of heat absorbed changes a particle of water into the vapor we call steam. This change is accompanied by an immense increase in volume, and as the boiler imprisons the steam and exactly limits the space it may occupy, each new particle thus changed crowds on those gone before and the imperative tendency to occupy more space begets the expansive force or pressure of steam which our gage registers. To hold this pressure with safety, is the *second office* of the boiler. If there be just room in the boiler above the water line, to contain one pound of water converted into steam at atmospheric pressure, the second pound thus converted *crowds* the first into *half this space*, appropriates the other half itself and thereby adds fully fifteen pounds per square inch to the *originally existing pressure*, and so on with each succeeding pound of water which the heat absorbed changes into steam. At the same time each pound of water previously converted into steam must absorb a certain quantity of heat to enable it *to retain its gaseous form* under this increased pressure, or some portion of it will fall back as watery spray. Every one who has seen a teakettle boil knows that the steam rises in transparent bubbles, which burst as they reach the surface, scattering spray to all sides but mainly upwards. The *spray*, being water, has no expansive force, and when allowed to leave the boiler with the steam not only represents so much *inert matter* carried along but presents innumerable surfaces to invite and hasten condensation. The *third office* of a good boiler is therefore the *separation* of this entrained water from the *steam*. This is an important office and worthy of the serious thought of the designer; yet it is often neglected in superstitious reliance on the fetich of an excessive amount of heating surface.

The water with which boilers are fed is rarely even approximately pure. Salts of lime and magnesia are the most frequent *impurities chemically combined*, while much extraneous matter both vegetable and mineral is carried along mechanically. The latter as well as the *carbonates* are *readily precipitated* at the boiling point at atmospheric pressure. But the *sulphates of lime and magnesia* require a temperature of nearly 300° Fahrenheit to become insoluble and drop to the bottom; this is about the boiling point for water under fifty-two pounds gage pressure. While therefore the common exhaust feed water heater and the old time mud drum will, if properly proportioned to the work remove the mud and the carbonates, they will have no effect whatever on the sulphates. For it is matter of common experience that you can almost hold your hand on the mud drum of a battery of boilers while they are under 100 pounds of steam, especially where the old method of feeding through the mud drum is adhered to, and an exhaust feed heater cannot yield more than 212° Fahrenheit temperature. The *sulphates make the hardest scale* when allowed to bake on the heating surfaces. Their removal is therefore even more necessary than that of the mud or the carbonates. If a mud drum or other vessel is made part of the boiler for this purpose it must be placed where it will necessarily *partake of* or *approximate* the *steam temperature*.. The *best modern practice removes all these impurities* by live steam purifiers, by chemical precipitation, or by filtration

—129—

after coagulation, *before* feeding the water to the boilers. But this best practice is not as yet the general rule, and these means may sometimes prove inadequate. Therefore a good boiler should be able to dispense with them, or, when supplied, to supplement their work.

The *fourth* office of the boiler is then to remove all impurities from the water which may have escaped other cleaning agencies, and to deposit them at points where they do the least harm and can be readily removed. No means are so efficient for this purpose as *positive and unchecked circulation* through all parts of the boiler, to keep the heating surfaces swept clean; and the vessel to catch the impurities must be open to the main current. If it can be arranged so as to precipitate most of the foreign matter out of the water *before it enters* into the main *circulation* the result will be still better.

The first office of the boiler, the absorption of the furnace heat and its transmission to the water requires *thin and homogeneous metal* for the heating surfaces and a strong and positive circulation of the water. It is well known that *a tube or flue* has much *greater strength* against *internal* than against *external* pressure. It is much easier to produce and maintain circulation through a tube than round about it. Finally it is much easier to clean the inside of tubes thoroughly, than the outside when they are grouped close together in a boiler. An iron tube of standard gage will stand 2,500 pounds to the square inch of internal pressure before rupture, and the rupture in the vast majority of cases is small and local. The same tube would collapse under external pressure much earlier, and once begun the collapse would be practically total.

Mr. Thomas Craddock of England, found by experiment that a *velocity of water* two miles per hour over tube heating surface *doubled its efficiency* in heat absorption, and that this circulation became more important the less the difference in temperature between the heat giving and the heat receiving body. Therefore in the ultimate economy of a boiler, to realize all the heat possible from the escaping temperature of the gases, circulation is all important. The *water tube* then best fulfills the first, second and fourth offices above explained, and must therefore become a *fundamental element* of the *Modern Boiler*. It is evident that for the third office, the separation of the entrained water from the steam, another element must be added to the water tubes. With few exceptions water tube boilers are supplied with a large drum or several drums or shells for this purpose. Observation of the boiling of water in an open vessel shows that the spray will, as the steam bubbles burst, fly upwards a number of inches. There is reason to believe that in a closed vessel under pressure it will not fly quite so far, certainly not further. Steam at 100 pounds gage pressure is about seven times as heavy as at atmospheric pressure, and hence occupies only one-seventh of the space. The *same weight of water* evaporated per second under the *higher pressure*, will rise to the surface in much smaller bubbles, or in a smaller number, or most probably both. The speed with which the steam rises through the water depends on the difference between the weight of the steam and that of the water. At atmospheric pressure the water weighs 1,570 times as much, at 100 pounds gage pressure 213 times as much as the steam. For these two reasons then the speed and energy with which the high pressure steam rises will be much less than that observed at atmospheric pressure. Under normal conditions therefore there is *less danger of priming* or wet steam at *high pressures* than at lower ones. But if by accident or design a large valve be suddenly opened much entrainment follows. This is because the sudden lowering of the pressure in the boiler temporarily increases the *rate of evaporation* enormously. This accounts for the *geyser like action* of certain boilers, mainly of a vertical type, which just previously have been working "like a charm," as soon as a sudden demand causes the engine valve to reach out for full stroke steam. From the above explanations it is evident that a reasonable height of steam space and a large surface at the water line will prevent priming under ordinary conditions, and some form of dry pipe placed well above the water line will take care of moderate fluctuations. If we can further so direct the circulation that the film of each bursting bubble is thrown *in a direction contrary to the steam delivery*, we will have a living active force to counteract any rush of spray towards the steam nozzle. As these arrangements can most readily be made in a *water tube boiler*, this then best fulfills the *third office* of a good modern boiler, the *separation of the entrained water* from the steam.

Compare for a moment the favorite type of fire tube boiler, the horizontal multitubular. Following the demands for a large heating surface, the tubes are crowded in close together and above the center of the shell, leaving only about *one-fifth of its area* as steam space, whose height is about one-fourth of the diameter. A recent report (A. B. M. A. 1892) shows that this tendency has gone so far that 30 per cent more tubes are put into boilers than the best rules for tube-spacing (A. B. M. A. 1889) warrant. This means that the *steam space* and the *steam liberating surface* have been much encroached on. Not only is the water line brought up too near the steam nozzle, but the channel for the rising steam bubbles is so curtailed and cut up that they create great commotion at the water line, and increase the tendency to prime. The *upper surface* of the water is generally accepted as the *steam liberating surface*. If all the steam were made on the surface of the upper row of tubes this would be correct. But all that is made on the bottom and sides of the shell, and on all the tubes below the top row has to pass the *narrow spaces between the tubes* of the upper rows. These are frequently but little over an inch wide, and have to serve for the return circulation of the water as well as the upward rush of steam mingled with water. Mr. Geo. H. Babcock, M. E., in a very instructive lecture on the circulation of water delivered at Cornell in 1890, suggests an ingenious method of approximately finding the speed of such rising currents. In a 60-inch boiler it would probably not be far from fourteen feet per second or say about ten miles an hour. Water rushing at ten miles an hour through a narrow slit will do a *good deal of sputtering*, and when it is half steam it will be practically all spray. The four or five inch body of water over the top row of tubes has a slight retarding influence but the *real liberating surface* for the steam is nevertheless the aggregate of the *narrow spaces between the upper tubes*. Where there is any scale or mud present in the water, its location and appearance after a fortnight's run shows that the bulk of the upward circulation in a horizontal tubular boiler is confined to a short section near the bridge wall, its speed decreasing towards front and rear till it meets the downward currents which are strongest near the ends of the boiler. This further concentrates the steam delivery on a small portion of the liberating surface. For this reason this whole type of fire tube boilers gives wet steam when forced. This has lead to insistence on more heating surface, and this again when supplied without due increase in the other important ratios of tube spacing, liberating surface and steam room, *serves*, as we have seen, *to increase the evils it is intended to remedy*. It must of course be conceded that in the boilers of the water tube type with either tubes or drums placed vertically or nearly so, the tendency to prime is even greater than in the horizontal fire tube types. But in the types which *have stood the test of years* the tubes and shells or drums are *horizontal* or *slightly inclined*, *full; half the shell is steam space*, the *vertical distance* from water line to steam nozzle is *half the diameter* or more, the upward *current of circulation is deflected* away from the steam opening, and the *liberating surface* is the *largest horizontal section* of the shell, entirely *free* from tubes or other obstructions. Well designed boilers of this class have been forced to nearly double their rated capacity without approaching the amount of entrainment considered permissible in the horizontal tubular type at conservative rating.

As these advantages are obtained with shells or drums of about half the diameter of fire tube boilers of the same evaporative capacity, *greater safety at high pressures* is the result. For the thinner metal has more strength per sq. in., and uniformity than thicker plate of the same quality. The rivet seams admit of more favorable proportions. Thin sheets can be better fitted than thick ones, etc. Thin metal transmits heat more rapidly than thick, and hence suffers less deterioration, and finally the *nest of tubes* in a water tube boiler *protects the shell* from the direct and fiercest heat, thus ensuring greater durability, and removing all danger of any chemical action of the hot carbon or sulphur on the steel boiler plates. The free circulation in a water tube boiler tends to equalize the temperatures all over the structure, thus *preventing* those *dangerous strains* due to *unequal expansion*. The old saw of "ice at the bottom, water in the middle, and steam on top" is but a slight exaggeration of what often occurs in a fire tube boiler, and many a "mysterious" explosion may be due to such a cause. These are some of the points of superiority of the boiler proper. In relation to furnace and chimney there are several more.

In a *firetube boiler* the aggregate tube area *limits* the capacity of the *furnace*, and *checks* the work of the *chimney*. The cogent reasons against increasing it have been pointed out above. In a *water tube boiler* the flue areas can be *freely proportioned* to furnace and chimney and can even be *adjusted* to suit local conditions after the boiler is built and set, without disarranging any important ratios.

It is well known that ashes and soot soon *cut down* both *heating surface* and *flue area* in *fire tube* boilers, and that flame entering a tube is soon *extinguished;* careful experiments have shown "that the *quantities of water* evaporated by consecutive equal lengths of flue-tubes *decrease in geometrical progression.*" (D. K. Clark.)

In *water tube* boilers the ashes and soot find much less chance for lodgment, all the *heating surfaces* are constantly accessible, during service, for *inspection* and *cleaning;* the *flame* is constantly *regenerated* since in impinging against successive water tubes effete combinations are broken up and new ones formed; ocular demonstration of these facts is daily possible.

Finally, it is possible to concentrate more power in a single water tube boiler than in any of the fire tube types. Therefore considerations of *safety, durability, economy, space* and *accessibility* point to the *Water Tube Boiler* as naturally the basis of a modern boiler plant.

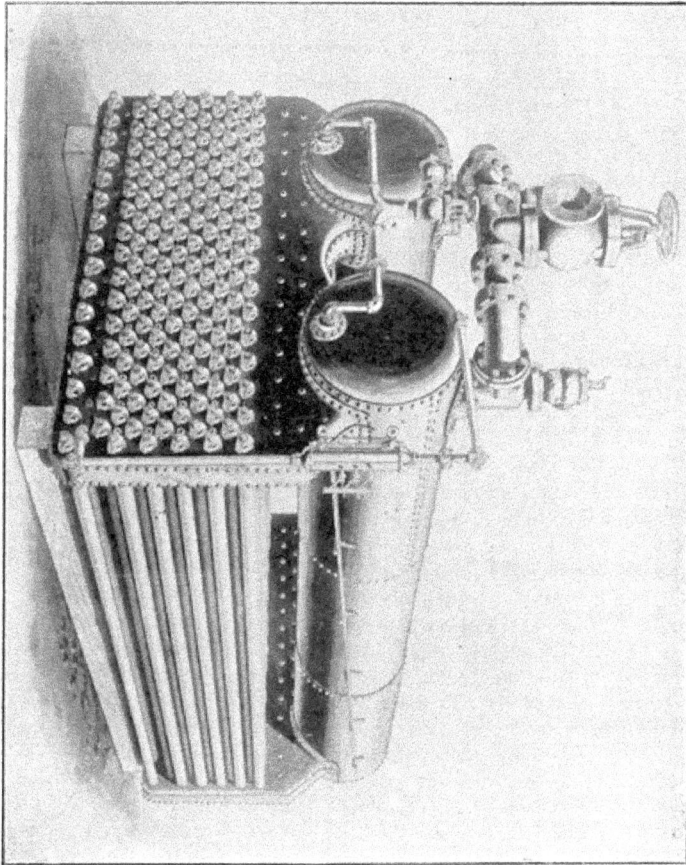

375 H. P. Heine Boiler.

DESCRIPTION OF THE HEINE SAFETY BOILER.

The boiler is composed of the best lap welded wrought iron tubes, extending between and connecting the inside faces of two "water legs" which form the end connections between these tubes and a combined steam and water drum or "shell," placed above and parallel with them. (Boilers over 200-horse power have two such shells.) These end chambers are of approximately rectangular shape, drawn in at top to fit the curvature of the shells. Each is composed of a *head plate* and a *tube sheet*, flanged all around and joined at bottom and sides by a butt strap of same material, strongly riveted to both. The water legs are further stayed by *hollow stay bolts* of hydraulic tubing, of large diameter, so placed that two stays support each tube and hand hole and are subjected to only very slight strain. Being made of heavy metal they form the strongest parts of the boiler and its natural supports. The WATER LEGS are joined to the shell by *flanged and riveted joints* and the drum is cut away at these two points to make connection with inside of water leg, the opening thus made being strengthened by bridges and special stays, so as to preserve the original strength.

THE SHELLS are cylinders with *heads dished* to form parts of a *true sphere.* The sphere is every where as strong as the circle seam of the cylinder which is well known to be twice as strong as its side seam. Therefore these heads require no stays. Both the cylinder and its spherical heads are therefore *free to follow* their *natural lines of expansion* when put under pressure. Where flat heads have to be braced to the sides of the shell, both suffer local distortions where the feet of the braces are riveted to them, making the calculations of their strength fallacious. This *we avoid entirely* by the dished heads. To the bottom of the front head a flange is riveted into which the feed pipe is screwed. This pipe is shown in the cut with *angle valve* and *check valve* attached.

On top of shell near the front end is riveted a *steam nozzle* or saddle, to which is bolted a Tee. This Tee carries the *steam valve* on its branch, which is made to look either to front, rear, right or left; on its top the *Safety Valve* is placed. The saddle has an area equal to that of Stop Valve and Safety Valve combined. The rear head carries a *blow-off flange* of about same size as the feed flange, and a *Manhead* curved to fit the head, the manhole supported by a strengthening ring outside. On each side of the shell a square bar, the *tile-bar*, rests loosely in flat hooks riveted to the shell. This bar supports the *side tiles* whose other ends rest on the *side walls*, thus closing in the furnace or flue on top. The top of the tile bar is two inches below *low water line.* The bars rise from front to rear at the rate of one inch in twelve. When the boiler is set, they must be exactly level, the whole boiler being then on an incline, i. e., with a fall of one inch in twelve from front to rear.

It will be noted that this makes the height of the *steam space* in front about *two-thirds* the diameter of the shell, while at the rear the *water* occupies *two-thirds* of the shell, the whole contents of the drum being equally divided between steam and water. The importance of this will be explained hereafter.

THE TUBES extend through the tube sheets into which they are expanded with roller expanders; opposite the end of each and in the head plates

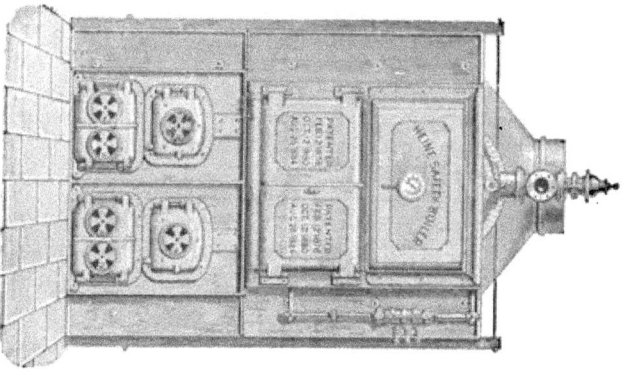

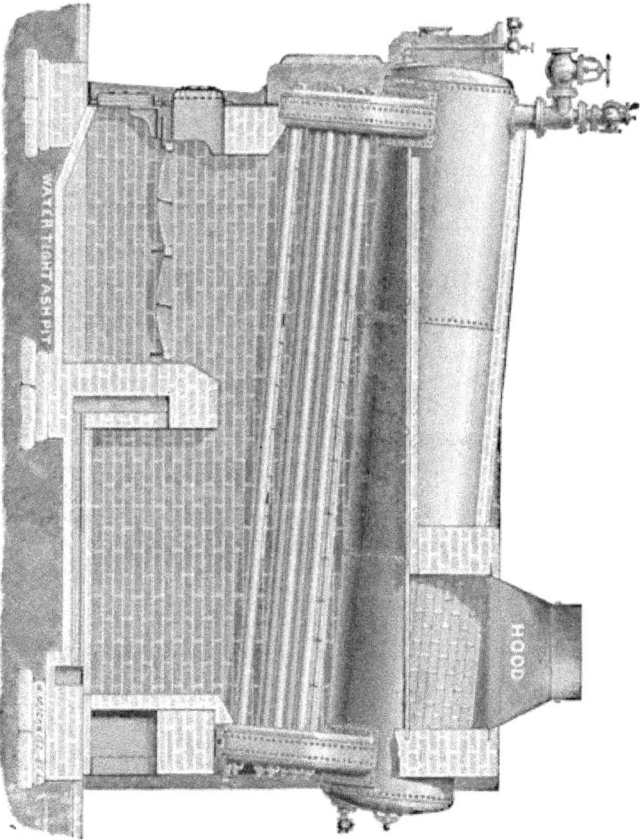

The Heine Safety Boiler.

is placed a hand hole of slightly larger diameter than the tube and through which it can be withdrawn. These hand holes are closed by small cast iron *hand hole plates*, which by an ingenious device for locking can be removed in a few seconds to inspect or clean a tube. The cut opposite shows these hand hole plates marked H. In the upper corner one is shown in detail, H₂ being the top view, H₃ the side view of the plate itself, the shoulder showing the place for the gasket. H₁ is the *yoke* or *crab* placed outside to support the bolt and nut.

Inside of the shell is located the *mud drum* D, placed well below the water line usually paralled to and three inches above the bottom of the shell. It is thus *completely immersed* in the hottest water in the boiler. It is of oval section slightly smaller than the manhole, made of strong sheet iron with cast iron heads. It is entirely enclosed except about eighteen inches of its upper portion at the forward end, which is cut away nearly parallel to the water line. Its action will be explained below. The *feed pipe* F enters it through a loose joint in front; the *blow-off pipe* N is screwed tightly into its rear head, and passes by a steam tight joint through the rear head of the shell. Just under the steam nozzle is placed a *dry pan* or *dry pipe* A. A *deflection plate* L extends from the front head of the shell inclined upwards, to some distance beyond the mouth or throat of the front water leg. It will be noted that the throat of each water leg is large enough to be the practical equivalent of the total tube area, and that just where it joins the shell it increases gradually in width by double the radius of the flange.

ERECTION AND WALLING IN.

In setting the boiler we place its front water leg firmly on a set of strong cast iron columns, bolted and braced together by the door frames, deadplate, etc., and forming the fire front. This is the fixed end. The rear water leg rests on *rollers* which are free to move on *cast iron plates* firmly set in the masonry of the low and solid rear wall. Wherever the brickwork closes in to the boiler broad joints are left which are filled in with tow or waste saturated with fireclay, or other refractory but pliable material. Thus *the boiler and its walls* are each *free to move separately* during expansion or contraction, without loosening any joints in the masonry. On the lower, and between the upper tubes, are placed light fire brick tiles. *The lower tier* extends from the front water leg to within a few feet of the rear one, leaving there an upward passage across the rear ends of the tubes for the flame, etc. *The upper tier* closes in to the rear water leg and extends forward to within a few feet of the front one, thus leaving the opening for the gases in front. *The side tiles* extend from side walls to tile bars and close up to the front water leg and front wall, and leave open the final uptake for the waste gases over the back part of the shell, which is here covered above water line with a row lock of *firebrick* resting on the *tile bars*. The rear wall of the setting and one parallel to it arched over the shell a few feet forward form the uptakes. On these and the rear portion of the side walls is placed a light sheet-iron hood, from which the breeching leads to the chimney. When an iron stack is used this hood is stiffened by L and T irons so that it becomes a truss *carrying the weight* of such stack and distributing it to the side walls. A good example of this latter style of braced hood is seen in the half tone cut of the *People's Railway Co.*, on page 51, where the four side walls of the three 200 horse-power boilers thus carry the heavy stack. In the *Central Distillery Plant*, (see half tone cut

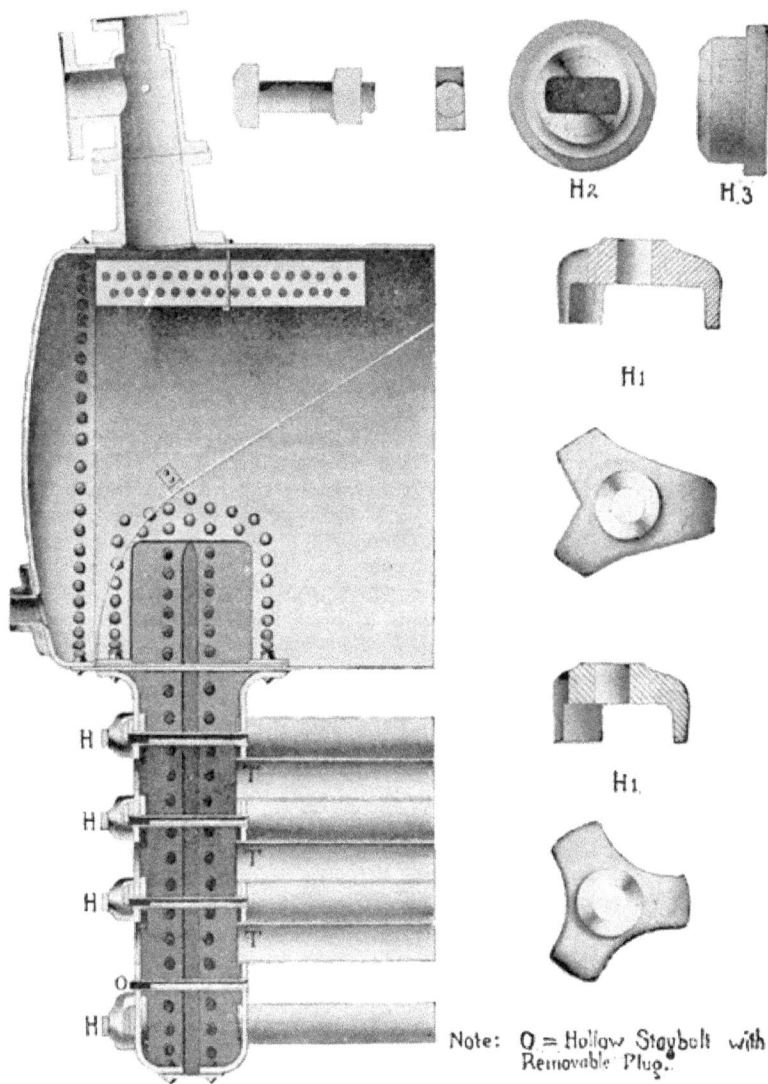

H_2

H_3

H_1

H_1

Note: O = Hollow Staybolt with Removable Plug.

Detail of Water-leg. Hand Hole Plates and Yokes, etc., of Heine Boilers.

on page 120, three of the 300 horse-power boilers are thus equipped, while the fourth boiler, put in later, carries its stack in the same way. In the *Union Depot Ry. Plant*, 1750 horse-power (see half tone cut on page 128), the hood is dispensed with and a long breeching, circle top, flat bottom, runs over all the boilers, its width spanning the distance between uptake walls; over each boiler is placed a stout cast iron frame, bolted to the bottom of the breeching and containing a swinging damper. The *Anheuser-Busch Plant*, 2400 horse-power (see half tone cut on page 170 has a circular iron flue supported on I beams just over the rear aisle, into which short necks from the hoods open from the side ; each neck contains a swinging damper. We are often obliged by local circumstances to carry the breeching out forward or midway of the boiler to one side. There is no difficulty of adapting our flue connections to such conditions. *Swinging dampers* are always to be *preferred;* sliding dampers are apt to stick, and always require considerable force to move them. The cut on page 139 shows the style of setting generally used by us. With moderate firing and dry coals, it will practically prevent smoke. With highly bituminous coals and somewhat pushing the fires some smoke will result. The bridge wall is hollow and has small slotted openings in rear to deliver hot air into the half consumed gases which roll over the bridge wall into the combustion chamber. It receives its air from channels in the hollow side walls (controlled by small cast iron slides), through a cross flue at the rear end and a number of small flues under the floor of the combustion chamber, as shown in the cut. In the rear wall of the combustion chamber is an arched opening, closed by a cast iron door, which in turn is shielded by a dry firebrick wall easily removable. For special fuels, for smoke prevention, etc., there are now to be had various forms of furnaces, automatic stokers, rocking grate-bars, etc. Heine boilers have been set and operated successfully with these various devices. They are not all equally applicable in all localities nor adapted to the same conditions. As a rule we find that our customers or their engineers understand their local fuels and local conditions best, and we are always glad to adapt our setting to such of these devices as they may select.

OPERATION.

The boiler being *filled to middle water line*, the fire is started on the grate. The flame and gases pass over the bridge wall and under the lower tier of tiling, finding in the *ample combustion chamber*, space, temperature and air supply for complete combustion, *before* bringing the heat in contact with the main body of the tubes. Then, when at its best, it rises through the spaces between the rear ends of the tubes, between rear waterleg and back end of tiling, and is allowed to expend itself on *the entire tube heating surface without meeting any obstruction.* Ample space makes leisurely progress for the flames, which meet in turn all the tubes, lap round them and finally reach the second uptake at the forward end of the top tier of tiling with their temperature reduced to less than 900° Fahrenheit. This has been measured here, while *wrought iron would melt* just above the lower tubes at rear end, showing a reduction of temperature of over 1,800° Fahrenheit between the two points. As this space is studded with water tubes swept clean by a positive and rapid circulation, the absorption of this great amount of heat is explained. The gases next travel under the bottom and sides of the shell and reach the uptake at just the proper temperature to produce the draft required. This varies of course according to

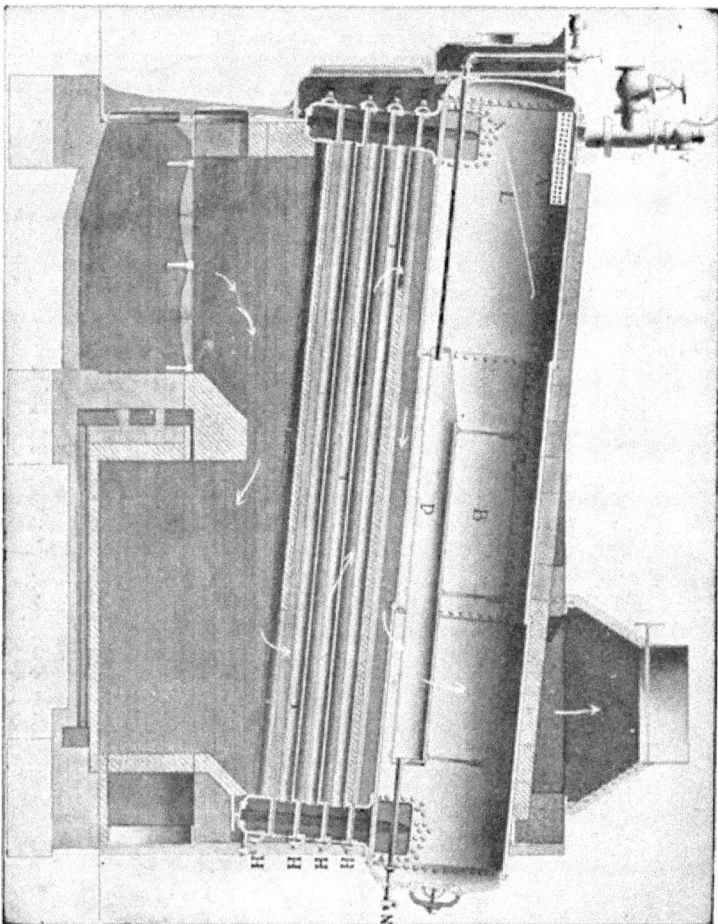

Longitudinal Section of Heine Boiler.

chimney, fuel, duty required, etc. With boilers running at their rated capacity 450° Fahrenheit are seldom exceeded. Meanwhile *as soon as the heat strikes the tubes* the circulation of the water begins. The water nearest the surface of the tubes becoming warmer rises, and as the tubes are higher in front this water flows towards the front water leg where it rises into the shell, while colder water from the shell falls down the rear water leg to replace that flowing forward and upward through the tubes.

This circulation, at first slow, *increases in speed* as soon as steam begins to form. Then the speed with which the mingled current of steam and water rises in the forward water leg will depend on the difference in weight of this mixture, and the solid and slightly colder water falling down the rear water leg. The cause of its motion is exactly the same as that which produces draft in a chimney as explained in the discussion of *"A Modern Boiler Plant,"* page 116. The maximum velocity will be reached when the mixture is about half steam and half water. As the area of the throat of the water leg is *practically equivalent* to the *aggregate tube area* (offsetting the greater amount of skin friction in the tubes against the reduced area of the throat), there will be nothing to interfere with the *free action of gravity and the full speed will be maintained* as long as steam is being made. This circulation must be well borne in mind. It is forward through the tubes, upward through the front water leg, to the rear in the shell, and down through the rear water leg. At the forward throat of the shell the channel slightly enlarges by reason of two outward flanges of the water leg. This *greatly facilitates the liberation of the steam,* and is the best form of orifice. (Bateman's experiments, Proc. Inst. Mech. Eng'rs, 1866, gives this form of orifice 95 per cent of theoretical capacity.) The deflection plate L assists in directing the circulation of the water to the rear. Thus the steam bubbles obtain a trend towards the rear, throwing the spray in a direction away from the flow of steam. It also has the effect of *increasing the liberating surface.* For each section of this moving surface of water, as it is delivering its load of steam, sweeps rapidly to the rear, making room for the next section, thus constantly presenting a fresh surface for this work.

The shallowness of the water at the front of the shell makes it easier for the steam to pass through; its depth at the rear ensures *a solid body of water* for replenishing the rear water leg and tubes. The height of the steam space in front removes the nozzle far out of reach of any spray; the deflection plate catches and deflects any sudden spurt, while finally the dry pan or dry pipe draws the steam from a large area, from three sides, *thus preventing any local disturbance.* These appliances make it possible to run the Heine Boiler 50 per cent above rating with *less than one-fifth of one per cent entrainment.*

The action of the *mud drum* is as follows: The feed water enters it through the pipe F about one-half inch above its bottom; even if it has previously passed the best heaters it is colder than the water in the boiler. Hence it drops to the bottom, and, impelled by the pump or injector, passes at a *greatly reduced speed* to the rear of the mud drum. As it is gradually heated to near boiler temperature it rises and flows slowly in reverse direction to the open front of the mud drum; here it passes over in *a thin sheet* and is immediately swept backward into the main body of water by the swift circulation, thus becoming *thoroughly mixed* with it before it reaches the tubes. During this process the mud, lime salts and other precipitates are deposited as a sort of semi-fluid "sludge" near the rear end of the mud drum, whence it is blown off at frequent intervals through the blow-off valve N. As the speed in the mud drum is only about one-fiftieth of that in the feed water pipe, plenty of time is given for this action. Any precipitates which may escape the mud drum at first, will of course form a scale on the inside of the tubes, etc. But the action of expansion and contraction cracks off scale on the *inside* of a tube *much faster* than on the *outside*, and then the circulation sweeps the small chips, like broken eggshells, upward, and as they pass over the mouth of the mud drum *they drop in the eddy*, lose velocity in this slow current and fall to the bottom, and, being pushed by the feed current to the rear end, are blown off from the mud drum with other refuse. On opening a Heine boiler after some months service, such bits of scale, whose shape identifies them, are always found in the mud of the mud drum. *Very little loose scale* is found on the bottom of the water legs; the current through the lower tubes, always the swiftest, brushes too near the bottom to allow much to lodge there.

This explanation of the action of the mud drum shows how *the inside of the tubes may be kept clean.* To keep the *outside clear of soot and ashes* which deposit on, and sometimes even bake fast to the tubes, each boiler is provided with two special nozzles with both side and front outlets, a short one for the rear, a long one for the front. They are of three-eighth inch gas pipe and each is supplied with steam by a one-half inch steam hose. The nozzle is passed through each stay bolt in turn, and thus delivers its side jets on the three or four tubes adjacent, with the full force of the steam, at the short range of two inches, *knocking the soot and ashes* off completely, while the end jet carries them into the main draft current to lodge at points in breeching or chimney base convenient for their ultimate removal. An inspection of the cuts will show that the stay bolts are so located that the nozzle can in turn be brought to bear on all sides of the tubes. As soon as the nozzle is withdrawn from the stay bolt, this is closed air-tight by a plain wooden plug.

In cleaning a boiler it is only necessary to *remove every fourth or fifth handhole plate* in the front water leg; the water hose, supplied with a short nozzle, can be entered in all the adjacent tubes, owing to the ample dimensions of the water leg. In the rear water leg only one or two handholes in the lower row need be opened to let the water and debris escape. The others in rear water leg are frequently *left untouched for years.* A lamp or candle hung on a wire through the manhead may be held opposite each tube so that it can be perfectly inspected from the front. Once or twice a year, where *the water is very scale bearing*, it may be advisable to take off all the handhole plates of the front water leg and pass a scraper through all the tubes in succession. Aside from the plain cylinder boiler there is no boiler so completely accessible for internal and external inspection as the Heine. The ashes which deposit in the combustion chamber are removed through the ashpit door in the rear wall, never allowing it to become more than one-third full.

We furnish with each boiler a set of "Rules for operation" in a neat frame, adapted to be hung up in the boiler room.

SUPERIORITY OF THE HEINE SAFETY BOILER.

In the discussion of A Modern Steam Plant we have pointed out the four principal offices of a good boiler, and have explained why water tube boilers best fulfill the conditions of the problem. Without denying the merits of other systems of construction, we claim that the Heine boiler stands at the very head and front in the good qualities essential to complete performance.

    1st.   It best absorbs and transmits heat; hence economy and capacity.

    2d.   It will hold high pressures with greatest safety.

    3d.   It best separates the Steam from the Water, ensuring Dryness.

    4th.   It is best adapted to precipitate and discharge scale and mud.

We ask a fair and critical examination of our description of the Heine Boiler, to which we shall refer in elucidating the above points.

ABSORPTION AND TRANSMISSION OF HEAT.

This, the most important work of the boiler, determines its economy and capacity, and must be discussed in connection with the furnace and the draft. For it is not sufficient to so construct the boiler that it will best absorb and transmit the heat, but it must also be so arranged that the heat can best reach it, and that nothing in its design will interfere with the best plan of furnace construction, nor increase unnecessarily the demands on the chimney.

For absorbing and transmitting heat nothing can be better than a nest of tubes placed entirely in the flue, which the hot products of combustion must traverse on their way from combustion chamber to chimney, especially when free and unimpeded circulation of the water is provided for. Mr. Babcock, in his interesting lecture on water circulation (Cornell University, 1890), has shown with great clearness that it depends, not as some have supposed, on the amount of inclination of the tubes, but "is a function of the difference in density of the two columns," the one of mingled steam and water, the other of solid water. The simple mode of calculation he suggests for finding the velocity of circulation gives us about twelve to eighteen feet as the average natural speeds for that general class of water tube boilers of which the Heine is a type. The cause of the circulation once understood, it is clear that any sharp turns or contractions which offer resistance to the flow will retard it in two ways. First, by altering the conditions of equilibrium on which the speed depends. Second, since a river can not rise higher than its source the speed lost by such an obstacle can not be regained; the loss in speed at this point will therefore be multiplied, at other points having larger areas, by the ratio those areas bear to this contracted one. In most boilers of this class there are between the tubes and the drum several points where the contents of seven, nine or even twelve tubes have to pass through an opening equal to one tube area. Every such place first disturbs the conditions on which the speed depends by absorbing some of the existing "head" (or difference in weight). Second, the maximum speed depending on the head can exist only at the least such opening, and hence in the nest of tubes the circulation will be reduced to one-seventh, one-ninth, or one-twelfth of the natural speed. In Heine Boilers there are no such contractions of area, even the smallest throat areas being 65 to 90 per cent of the aggregate tube area.

The Heine Boiler gains another advantage from this fact. The water in the upper tubes having less "head," begins with less speed than that in the lower tier; the heating surface of the upper tubes will then be somewhat less active than that of the lower tubes. Since they get the first heat, more steam will be made in the lower tubes, further increasing the original difference in velocity. The combined effect is that the circulation through the lower tubes is much faster than through the upper ones. The obstructions before noted will multiply this difference, since only the more rapid current will there make its way at the expense of the sluggish one. Thus the effectiveness of the upper tubes is largely curtailed. The full throat area of the Heine Boiler, on the other hand, leaves room for all the currents, hence the full efficiency of the upper tubes is preserved.

In the older types of this class of water tube boilers the tubes only are inclined, and therefore the return circulation in the rear has to pass through small tubes several feet in length, nearly vertical. The escaping gases pass around them, tending to create an upward circulation along the surface, which must somewhat check the downward flow. Everybody daily observes that water invariably "swirls" when it escapes through a small round hole or a tube from a wash bowl, bath tub or barrel. We all know how vexatious is the delay caused by it. This action, being independent of the surrounding pressure, takes place in the short tubes just mentioned, and retards the flow.

In the Heine boiler this is done away with. The water at the rear end of the shell is about a foot deeper than in front, the openings are large and rectangular, and the downward flow is through a rectangular chamber equal in section to the aggregate tube area. Swirling is impossible and the tubes are fully supplied with solid water under all circumstances.

The circulation of the water is the life of all water tube boilers. Craddock's experiments show how its speed multiplies the effectiveness of heating surface. Details of construction which reduce it to less than one-fifth its natural velocity are therefore faulty, especially when this reduced speed is found in the tubes. The Heine Boiler carefully avoids any such obstructions and the natural speed of circulation is maintained throughout.

Therefore the effectiveness of its heating surface for the absorption and transmission of heat is much greater than that of other boilers.

All fuels require much air, great heat, space for expansion, and time for their complete combustion. An arched chamber, composed entirely of fire brick, would be the ideal furnace, in which combustion should be completed without meeting any cooling surface, the products when at their greatest temperature to be launched into and amongst the heating surfaces of the boiler. The nearer a furnace can be made to approach these conditions the better will be its work. The other extreme is the internally fired

100 H. P. Heine Boiler over Puddling Furnace at United States Iron and Tin Plate Works, DEMMLER, PA.

boiler, whose performance on bituminous coals is very inferior in spite of its smaller loss by radiation. Between them lie the return tubular boilers, and those water tube boilers whose furnaces are separated from their combustion chambers by the first pass of the nest of tubes. The heating surfaces of a boiler are such for the water only; in reference to the flame they are cooling surfaces. Brought in contact with the gases at the be ginning of combustion they lower their temperature below the required point. This results in the direct loss of much of the heating power of the volatile part of the fuel which escapes unburnt, and in the indirect loss due to impairment of the conductivity of the heating surface owing to deposit of much soot. As the first third of the heating surface thus encountered absorbs between 60 and 70 per cent of the heat (Graham's experiments, 1858), it is useless to expect secondary combustion of any practical value in a combustion chamber placed beyond it, with no means of restoring the lost temperature. This method of construction probably grew out of the pretty widespread belief that heating surface placed at right angles to the course of the flame was much more effective than in any other relative position. Even if this were true the old adage, "always catch your hare before you cook him," should induce prudent men not to allow its application to vitiate their furnace construction. It is probably true only for radiant heat; no experiments are adduced to prove it true for currents of hot gas; there it is plainly a case of "faith without works." On the other hand German experiments (Stuehlen Ing. Kal., 1892) show tube heating surface parallel to the current 30 per cent more effective than when placed at right angles. The Heine boiler setting approximates the ideal furnace. Fire place and combustion chamber are of fire brick, except that minimum of tube surface required to support the fire brick roof, experience having shown that arches are too short-lived where the soda of the ashes under high temperatures fluxes the fire brick. The radiation from side walls and floor is arrested and utilized to pre-heat the small amount of air thrown into the gases at the bridge wall. Having passed the combustion chamber, flame and gases are thrown in contact with the whole of the tube heating surface, which they envelope and strike at all angles, the main trend being parallel to the tubes. Observation shows that they roll around, mix, break up, combine, etc., according to natural laws, and following many causes, to the apparent neglect of some single one the professor may lay down in the lecture room, or the draftsman prescribe by the conventional arrow. In the Heine boiler and furnace we arrange for space, time, air and heat for the best combustion, then open out into an ample flue, containing all the tubes, and like the Brooklyn alderman with the gondolas, "leave the rest to nature." The small tiles on the upper and lower tier of tubes make adjustments of flue areas, to suit local and possibly changing requirements, possible at all times. The trend of the gases is the natural one, rising gradually towards the stack. We thus avoid that loss in chimney power incident to pulling hot gases downwards against their bent.

Having shown that with the most free circulation of the water, we combine the best furnace arrangement, the natural circulation of the hot gases, the equal exposure of the total heating surface to them, and the least demands on the chimney, we have explained why the Heine Boiler ranks first in economy and capacity. Our many customers will gladly attest the results.

The facilities for observing and cleaning the heating surfaces through the hollow staybolts have been fully explained in the description of the boiler. The effect of this on the economy and capacity must be here noted. As human nature goes, the fireman will not begin to clean the heating surfaces until he has to. In the Heine boiler, as he blows through each staybolt in turn, the cleaned section and increased draft reward him at once by a rise in the steam pressure while cleaning. Under the old plan of cleaning through side doors in the walls, cold air rushes in, and the pressure drops while cleaning, and does not rise again until the work is completed and the doors again closed. Furthermore, the absence of these doors in the side walls of the Heine boiler makes them less liable to crack and leak.

SAFETY AT HIGH PRESSURES.

This depends on the qualities of the materials, the workmanship, the proper arrangement of the parts, avoidance of unequal expansion and contraction, and accessibility for inspection, cleaning and repairs.

We use no cast iron in any parts subject to tensile stress. In this we follow the rule laid down by the AMERICAN BOILER MANUFACTURER'S ASSOCIATION (Proceedings 1889):

CAST IRON—Should be of soft, gray texture and high degree of ductility. To be used only for hand-hole plates, crabs, yokes, etc., and manheads. It is a dangerous metal to be used in mud drums, legs, necks, headers, manhole rings, or any part of a boiler subject to tensile strains; its use should be prohibited for such parts."

For shells, water legs and drums we use a first-class flange steel made for us and inspected before it leaves the steel works under the following:

SPECIFICATIONS FOR BOILER PLATES FOR HEINE SAFETY BOILERS.

STEEL.—Homogeneous Steel made by the OPEN HEARTH process, and having the following qualities:

TENSILE STRENGTH.—55,000 to 62,000 lbs. per square inch.

ELASTIC LIMIT.—Not under 32,000 lbs. per square inch.

ELONGATION.—20 per cent for plates $\frac{5}{8}$ inch thick or less, 22½ per cent for plates over $\frac{5}{8}$ inch and under $\frac{3}{8}$ inch thick, 25 per cent for plates $\frac{3}{8}$ inch thick and over.

TEST SECTION.—To be 8 inches long, planed or milled edges; its cross sectional area shall not be less than one-half of one square inch, nor shall its width ever be less than the thickness of the plate. Every third test piece to be of the shape and dimensions prescribed by the rules of the United States Board of Supervising Inspectors of Steamboats.

BENDING TEST.—Steel up to ½ inch thickness must stand hot and cold bending double, and being hammered down on itself; above that thickness, it must bend round a mandrel of diameter one and one-half times the thickness of plate down to 180°. All without showing signs of distress.

NICKED SAMPLE.—When a sample is broken, after being nicked, the appearance of laminations or cold shuts, shall cause the rejection of the plates represented by the sample.

ALL TESTS.—To be made at the steel mill by the inspectors of the Robert W. Hunt & Co. Bureau of Inspection and Tests.

CHEMICAL TESTS—Will be required, and if they show more than 0.04 per cent Phosphorus, or more than 0.03 per cent Sulphur, the plates will be rejected.

This is the same as the standard adopted by the Americal Boiler Manufacturers' Association, except that we have increased the requirements for elongation somewhat; we have further added the tests on the section used by the United States Board of Supervising Inspectors, to meet the requirements of cities prescribing the "Marine" tests. It is well known that the same steel will show higher t. s. on the "Marine" section than on the 8 inch section, but the latter is best for showing the elongation.

The tubes are the standard American wrought iron boiler tubes, all tested by hydrostatic pressure at the tube mills. They are intended to be the weakest parts of the structure. As already explained, a tube giving way from internal pressure suffers a local rupture merely; the boiler will require several minutes to empty itself through a tube, resulting in a gradual though rapid decrease of pressure, an extinguishing of the fire, and no explosion.

The staybolts are made of best butt-welded hydraulic tubing. The threads on them are therefore cut into solid metal all around, which would be doubtful were lap-welded or built up tubing used. They are so proportioned that in testing to rupture they part in the solid metal but do not strip the thread. The ends are carefully peaned over.

The rivets are according to American Boiler Manufacturer's Association standard, which we quote:

"RIVETS to be made of good charcoal iron, or of a very soft, mild steel running between 50,000 and 60,000 pounds tensile strength and showing an elongation of not less than 30 per cent in eight inches, and having the same chemical composition as specified for plates."

In all the processes of manufacture we follow the best boiler shop practice of the United States as laid down by the American Boiler Manufacturers' Association, as for instance in the rule for flanging:

"FLANGING to be done at not less than a good, red heat. Not a single blow to be given after the plate is cooled down to less than cherry red by daylight. After flanging, all plates should be annealed by uniform cooling from an even dull red heat for the whole sheet in the open air."

Having built up our boiler of the very best materials, and by the best methods of workmanship, we erect it in such a way that there can be no unequal expansion strains.

The entirely free and unchecked circulation of water and fire has been fully explained; this equalizes temperatures not only when in full operation, but as soon as the fire is lit. This can be verified by feeling the ends of shell and water legs when starting fires. Besides this there is another equalizing tendency. The shell will **stretch more** than the tubes **from the internal pressure;** the lower tubes receiving **greater heat,** will **expand more from this cause.** The two tendencies counterbalance beautifully, as can be verified by delicate measurements on any Heine boiler while cold and while hot and under heavy pressure.

Our method of supporting the boiler on the water legs, the front one on a fixed support, the rear one on rollers, gives freedom for expansion without undue stress on any part. The **weight of the boiler filled with water** is thus carried on **its strongest parts.** Most sectional boilers can not be thus supported, having in place of the water legs, loose, many-jointed constructions incapable of supporting any extra weight.

It is evident that ours is a much better way to support a boiler than to hang it from a gallows frame by bolts or links. For these **concentrate** strains equal to the whole weight of boiler and water **on two points** of the shell, thus disturbing that equilibrium of stress obtained by giving it the cylindrical form. Another **signal advantage** of the Heine boiler is that it is **completed and thoroughly tested** in the boiler shop **before shipment.**

Our style of setting, with horizontal travel of the gases, has two further advantages over the up and down method.

1st. The **cold air** which rushes into the furnace when the doors are opened for firing is **drawn to the rear, away from the tube joints,** in place of up and among them.

2nd. The **hot gases** do not reach the shell until after passing the **entire tube heating surface,** being then no longer hot enough to injure a rivet joint; in the up and down type they make their **first turn under a rivet joint** of the shell, after traversing only a third of the tube surface, and in what is considered a combustion chamber **hot enough** to regenerate the flame. Hence our shells are safer!

In all water tube boilers access must be had to each tube through some form of hand hole plate. Some have each group of two, three or more tubes controlled by a hand hole plate, some each single tube. Of course the larger each such plate the more danger of cracks, leakage of joints, etc. Elsewhere we have explained why only a few hand hole plates of each set have to be removed for washing out a Heine boiler. But besides this our hand hole plates **are much safer** than others in general use. A typical form for sectional boilers is shown below. T T are the ends of the

Cast-Iron End Connection
Used on Sectional Boilers.

—148—

tubes and the joints are made **outside** as at J. J. on the cap C. On the inside is merely a yoke Y to hold up the bolt B. This of course necessitates another joint j under the nut. These joints have to be made tight **while the boiler is cold;** this requires a nice exercise of judgment, since strain enough must be put on the bolt both to counterbalance the internal pressure of the boiler when steam is raised, and enough more to keep the joint tight then. In other words, the stretch of the bolt has to be anticipated and more strain added. And this double strain is **always** on the bolt whether the boiler is under steam or idle. It will not do to tighten up on the bolt when the boiler is under steam. For leakage around the threads will soon fill the hollow cap of the nut, which at any additional turn of the nut will crack it open by hydrostatic pressure. If we have a hand hole of 4½ inches diameter we have an area of 15.9 square inches to cover. At 125 pounds steam pressure we have 1,987 pounds pressure under the cap and about 150 pounds more under the nut to counteract before any strain becomes available to make the joints tight. It has often happened that a cracked nut has caused a cap to blow off, scalding the attendants.

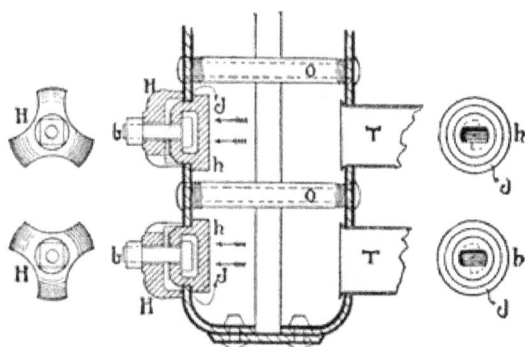

Plate Steel Water-Leg
of Heine Boiler.

With the Heine boiler the case is reversed; the single joint at J is an inside one, **this pressure of 1,987 pounds makes the joint,** so that the bolts can be drawn up when under steam, receiving but a trifling strain. It is clear that **this is the safe plan,** while the other is not. We have thus shown that in materials, workmanship, general design, settings, and in details of construction the Heine boiler is the safest.

SEPARATION OF WATER; DRYNESS OF STEAM.

In describing the functions of a boiler in a modern steam plant we have shown to what causes the entrainment of water is due. The description of the Heine boiler shows how the entirely unchecked **circulation tends away from** the steam nozzle. The steam bubbles, lighter than the water, pass through it on some diagonal course, a resultant from their own vertical trend and the backward flow of the water. This throws the spray away from the vapor with a momentum about two hundred times that of the steam which flows towards the nozzle, with about one-fourth of the speed it attains in the steam pipe. The function of the dry pipe or dry pan is well understood. Add this, and the action just described, to the fact that the inclination of our shell removes the water line further from the steam nozzle than in other boilers, and the reason why our steam is always dry becomes clear. An active agency for drying the steam, present at all times in the boiler, more vigorous the more the boiler is pushed, ensures dry steam always. On forcing tests we have shown steam six times as dry as our competitors. This has a decisive influence on the every day economy of a steam plant.

PRECIPITATION AND DISCHARGE OF SCALE AND MUD.

The Heine Boiler was originally developed under the difficult conditions of boiler practice in the great Mississippi Valley. The problem was not only the economic utilization of the highly bituminous coals, low in calorific value as they are high in ash and volatile matter, but also the making of steam from water strongly impregnated with mineral salts and frequently carrying a brown mixture of the sacred soils of several great States.

The faults of the old style of mud drum were here but too apparent. The various ingenious coil devices choked up the faster, the more effective they were. The "Spray Feeds" wet the steam in the exact ratio of their efficiency in scale precipitation. The Heine mud drum, holding the incoming feed water suspended for a time in an almost quiescent state, while subject to the external contact of a rapid current of the hottest water in the boiler, furnishes time, checked velocity and heat to induce precipitation. The necessity of a high temperature to make the mineral salts insoluble has been before explained. Evidence of it is found in every boiler. It is well known that any reduction in velocity favors the dropping of sediment. Instead of checking the speed of circulation in the tubes where the precipitates do harm, the Heine boiler provides this mud drum where no fire can get at them to bake them into scale, but where they can be collected and blown off at such intervals as their amount prescribes.

The fact that we have successfully replaced two-flue boilers in localities where return tubulars were tabooed on account of bad water proves the practical efficiency of our free circulation and submerged mud drum.

Chicago Athletic Club.
CHICAGO. ILL.
Contains 300 H. P. Heine Boilers.

Factors of Equivalent Evaporation from and at 212° F.

GAUGE PRESSURE, LBS. PER SQUARE INCH.

Temp. Feed.	40	50	60	70	80	90	100	110	120	130	140	150	160	170	180	190	200
32	1.211	1.214	1.217	1.219	1.222	1.224	1.227	1.229	1.231	1.233	1.234	1.236	1.237	1.238	1.239	1.240	1.241
35	1.208	1.211	1.214	1.216	1.219	1.221	1.224	1.226	1.228	1.230	1.231	1.233	1.234	1.235	1.236	1.237	1.238
40	1.203	1.206	1.209	1.211	1.214	1.216	1.219	1.221	1.223	1.224	1.226	1.229	1.229	1.230	1.231	1.232	1.233
45	1.197	1.200	1.203	1.205	1.208	1.210	1.213	1.216	1.217	1.219	1.220	1.222	1.223	1.224	1.225	1.226	1.227
50	1.192	1.195	1.198	1.200	1.203	1.205	1.208	1.210	1.212	1.214	1.215	1.217	1.218	1.219	1.220	1.221	1.222
55	1.187	1.190	1.193	1.195	1.198	1.200	1.203	1.205	1.207	1.209	1.210	1.212	1.213	1.211	1.215	1.216	1.217
60	1.182	1.185	1.188	1.190	1.193	1.195	1.198	1.200	1.202	1.204	1.205	1.207	1.208	1.209	1.210	1.211	1.212
65	1.177	1.180	1.183	1.185	1.188	1.190	1.193	1.195	1.197	1.198	1.200	1.202	1.203	1.204	1.205	1.206	1.207
70	1.172	1.175	1.178	1.180	1.183	1.185	1.188	1.190	1.192	1.193	1.195	1.196	1.198	1.199	1.200	1.201	1.202
75	1.167	1.170	1.173	1.175	1.178	1.180	1.183	1.184	1.187	1.188	1.190	1.191	1.193	1.194	1.196	1.196	1.197
80	1.161	1.164	1.167	1.169	1.172	1.174	1.177	1.179	1.181	1.183	1.184	1.186	1.187	1.188	1.189	1.190	1.191
85	1.156	1.159	1.162	1.164	1.167	1.169	1.172	1.174	1.176	1.178	1.179	1.181	1.182	1.183	1.184	1.185	1.186
90	1.151	1.154	1.157	1.159	1.162	1.164	1.167	1.169	1.171	1.173	1.174	1.176	1.177	1.178	1.179	1.180	1.181
95	1.146	1.149	1.152	1.154	1.157	1.159	1.162	1.164	1.166	1.168	1.169	1.171	1.172	1.173	1.174	1.175	1.176
100	1.141	1.144	1.147	1.149	1.152	1.154	1.157	1.158	1.161	1.162	1.164	1.165	1.167	1.168	1.169	1.170	1.171
105	1.135	1.138	1.141	1.143	1.146	1.148	1.151	1.153	1.155	1.157	1.158	1.160	1.161	1.162	1.163	1.164	1.165
110	1.130	1.133	1.136	1.138	1.141	1.143	1.146	1.148	1.150	1.152	1.153	1.155	1.156	1.157	1.158	1.159	1.160
115	1.125	1.128	1.131	1.133	1.136	1.138	1.141	1.143	1.145	1.147	1.148	1.150	1.151	1.152	1.153	1.154	1.155
120	1.120	1.123	1.126	1.128	1.131	1.133	1.136	1.138	1.140	1.141	1.143	1.145	1.146	1.147	1.148	1.149	1.150
125	1.115	1.118	1.121	1.123	1.126	1.128	1.131	1.133	1.135	1.136	1.138	1.139	1.141	1.142	1.143	1.144	1.145
130	1.110	1.113	1.115	1.117	1.120	1.122	1.125	1.127	1.129	1.131	1.132	1.134	1.135	1.136	1.137	1.138	1.139
135	1.104	1.107	1.110	1.112	1.115	1.117	1.120	1.122	1.124	1.126	1.127	1.129	1.130	1.131	1.132	1.133	1.134
140	1.099	1.102	1.105	1.107	1.110	1.112	1.115	1.117	1.119	1.121	1.122	1.124	1.125	1.126	1.127	1.128	1.129
145	1.094	1.097	1.100	1.109	1.105	1.107	1.110	1.112	1.114	1.115	1.117	1.119	1.120	1.121	1.122	1.123	1.124
150	1.089	1.092	1.095	1.097	1.100	1.102	1.103	1.107	1.109	1.110	1.112	1.113	1.115	1.116	1.117	1.118	1.119
155	1.083	1.086	1.089	1.091	1.094	1.096	1.099	1.101	1.103	1.105	1.106	1.108	1.109	1.110	1.111	1.112	1.113
160	1.078	1.081	1.084	1.086	1.089	1.091	1.094	1.096	1.098	1.100	1.101	1.103	1.104	1.105	1.106	1.107	1.108
165	1.073	1.076	1.079	1.081	1.084	1.086	1.089	1.091	1.093	1.091	1.096	1.098	1.099	1.100	1.101	1.102	1.103
170	1.068	1.071	1.071	1.076	1.079	1.081	1.084	1.086	1.088	1.089	1.091	1.092	1.094	1.095	1.096	1.097	1.098
175	1.063	1.066	1.069	1.071	1.074	1.076	1.079	1.081	1.083	1.084	1.086	1.087	1.089	1.090	1.091	1.092	1.093
180	1.057	1.060	1.063	1.065	1.068	1.070	1.073	1.075	1.077	1.079	1.080	1.082	1.083	1.084	1.085	1.086	1.087
185	1.052	1.055	1.058	1.060	1.063	1.065	1.068	1.070	1.072	1.074	1.075	1.077	1.078	1.079	1.080	1.081	1.082
190	1.047	1.050	1.053	1.055	1.058	1.060	1.063	1.065	1.067	1.068	1.070	1.072	1.073	1.074	1.075	1.076	1.077
195	1.042	1.045	1.048	1.050	1.053	1.055	1.058	1.060	1.062	1.063	1.065	1.066	1.068	1.069	1.070	1.071	1.072
200	1.037	1.040	1.043	1.045	1.048	1.050	1.053	1.055	1.057	1.058	1.060	1.061	1.063	1.064	1.065	1.066	1.067
205	1.032	1.035	1.038	1.040	1.043	1.045	1.048	1.050	1.052	1.053	1.055	1.056	1.058	1.059	1.060	1.061	1.062
210	1.026	1.029	1.032	1.035	1.037	1.040	1.042	1.044	1.046	1.047	1.049	1.051	1.052	1.054	1.054	1.057	1.058
212	1.024	1.027	1.030	1.033	1.035	1.038	1.040	1.042	1.044	1.045	1.047	1.049	1.050	1.052	1.054	1.055	1.056

TABLE NO. 71.

Diameters, Circumferences and Areas of Circles.

	Advancing by 10ths.								Advancing by 8ths.							
Diam.	Circum.	Area.	Diam.	Circum.	Area.	Diam.	Circum.	Area.	Diam.	Circum.	Area.	Diam.	Circum.	Area.	Diam.	Circum.	Area.

TABLE NO. 72.

Diameters and Circumferences of Circles, and the Contents in Gallons at One Foot in Depth.

Diameter. Ft.	Diameter. In.	Circum. Ft.	Circum. In.	Area in sq. feet.	Gallons. 1 Ft. Depth.
4		12	6¾	12.56	93.97
4	1	12	9⅞	13.09	97.93
4	2	13	1	13.63	101.97
4	3	13	4⅛	14.18	103.03
4	4	13	7¼	14.74	110.29
4	5	13	10½	15.32	114.57
4	6	14	1⅝	15.90	118.93
4	7	14	4⅝	16.49	123.38
4	8	14	7⅞	17.10	127.91
4	9	14	11	17.72	132.52
4	10	15	2⅛	18.34	137.21
4	11	15	5¼	18.98	142.05
5		15	8½	19.63	146.83
5	1	15	11⅝	20.29	151.77
5	2	16	2¾	20.96	156.78
5	3	16	5¾	21.64	161.88
5	4	16	9	22.34	167.06
5	5	17	0⅛	23.04	172.33
5	6	17	3¼	23.75	177.67
5	7	17	6⅜	24.48	183.09
5	8	17	9⅝	25.21	188.60
5	9	18	0¾	25.96	194.19
5	10	18	3⅞	26.72	199.86
5	11	18	7⅛	27.49	205.61
6		18	10⅛	28.27	211.44
6	3	19	7½	30.67	229.43
6	6	20	4⅞	33.18	248.15
6	9	21	2⅜	35.78	267.61
7		21	11⅞	38.48	287.80
7	3	22	9¼	41.28	308.72
7	6	23	6¾	44.17	330.38
7	9	24	4⅛	47.17	352.76
8		25	1½	50.26	375.90
8	3	25	11	53.45	399.76
8	6	26	8⅜	56.74	424.36
8	9	27	5¾	60.13	449.21
9		28	3¼	63.61	475.75
9	3	29	0⅝	67.20	502.55
9	6	29	10⅛	70.88	530.08
9	9	30	7½	74.66	558.35
10		31	5	78.54	587.35
10	3	32	2⅜	82.51	617.08
10	6	32	11¾	86.59	647.55
10	9	33	9¼	90.76	678.27
11		34	6⅝	95.03	710.69
11	3	35	4⅛	99.40	743.36
11	6	36	1½	103.86	776.77
11	9	36	10⅞	108.43	810.91
12		37	8⅜	113.09	848.18
12	3	38	5¾	117.85	881.39
12	6	39	3¼	122.71	917.73
12	9	40	0⅝	127.67	954.81
13		40	10	132.73	992.62
13	3	41	7½	137.88	1031·17

Diameter. Ft.	Diameter. In.	Circum. Ft.	Circum. In.	Area in feet.	Gallons. 1 Ft. Depth.
13	6	42	4⅞	143.13	1070.45
13	9	43	2¼	148.48	1108.06
14		43	11¾	153.93	1151.21
14	3	44	9⅛	159.48	1192.69
14	6	45	6⅝	165.13	1234.91
14	9	46	4	170.87	1277.86
15		47	1⅛	176.71	1321.54
15	3	47	10⅞	182.65	1365.96
15	6	48	8¼	188.69	1407.51
15	9	49	5¾	194.82	1457.00
16		50	3⅛	201.06	1503.62
16	3	51	0½	207.39	1550.97
16	6	51	10	213.82	1599.06
16	9	52	7⅝	220.35	1647.89
17		53	4⅞	226.98	1697.45
17	3	54	2⅛	233.70	1747.74
17	6	54	11⅜	240.52	1798.76
17	9	55	9⅛	247.45	1850.53
18		56	6½	254.46	1903.02
18	3	57	4	261.58	1956.25
18	6	58	1⅜	268.80	2010.21
18	9	58	10¾	276.11	2064.91
19		59	8¼	283.52	2120.34
19	3	60	5⅝	291.03	2176.51
19	6	61	3⅛	298.64	2233.29
19	9	62	0½	306.35	2291.04
20		62	9⅞	314.16	2349.41
20	3	63	7⅜	322.06	2408.51
20	6	64	4¾	330.06	2468.35
20	9	65	2¼	338.16	2528.92
21		65	11⅜	346.36	2590.22
21	3	66	9	354.65	2652.25
21	6	67	6½	363.05	2715.04
21	9	68	3⅞	371.54	2778.54
22		69	1⅜	380.13	2842.79
22	3	69	10¾	388.82	2907.76
22	6	70	8¼	397.60	2973.48
22	9	71	5⅝	406.49	3039.92
23		72	3	415.47	3107.10
23	3	73	0½	424.55	3175.01
23	6	73	9⅞	433.73	3243.65
23	9	74	7¼	443.01	3313.04
24		75	4¾	452.39	3383.15
24	3	76	2⅛	461.86	3454.00
24	6	76	11⅝	471.43	3525.59
24	9	77	9	481.10	3597.90
25		78	6⅜	490.87	3670.95
25	3	79	3⅞	500.74	3744.74
25	6	80	1¼	510.70	3819.26
25	9	80	10¾	520.76	3894.52

Wrought Iron, Steel, Copper and Brass Plates.

Birmingham Gauge.

No. of Gauge.	Thickness, Inches.	Weight Per Square Foot, Lbs.			
		Iron.	Steel.	Copper.	Brass.
0000	0.454 or 7/16 full........	18.2167	18.4596	20.5662	19.4312
000	0.425	17.0531	17.2805	19.2525	18.1900
00	0.38 or 3/8 full.........	15.2475	15.4508	17.2140	16.2640
0	0.34 or 1/3 full.........	13.6425	13.8244	15.4020	14.5520
1	0.3	12.0375	12.1980	13.5900	12.8400
2	0.284	11.3955	11.5474	12.8652	12.1552
3	0.259 or 1/4 full.........	10.3924	10.5309	11.7327	11.0852
4	0.238	9.5497	9.6771	10.7814	10.1864
5	0.22	8.8275	8.9452	9.9660	9.4160
6	0.203 or 1/5 full.........	8.1454	8.2540	9.1959	8.6884
7	0.18 or 3/16 light......	7.2225	7.3188	8.1540	7.7040
8	0.165 or 1/6 light......	6.6206	6.7089	7.4745	7.0620
9	0.148 or 1/7 full.........	5.9385	6.0177	6.7044	6.3344
10	0.134	5.3767	5.4484	6.0702	5.7352
11	0.12 or 1/8 light......	4.8150	4.8792	5.4360	5.1360
12	0.109	4.3736	4.4319	4.9377	4.6652
13	0.095 or 3/16 light......	3.8119	3.8627	4.3035	4.0660
14	0.083	3.3304	3.3748	3.7599	3.5524
15	0.072	2.8890	2.9275	3.2616	3.0816
16	0.065	2.6081	2.6429	2.9445	2.7820
17	0.058	2.3272	2.3583	2.6274	2.4824
18	0.049 or 1/20 light......	1.9661	1.9923	2.2197	2.0972
19	0.042	1.6852	1.7077	1.9026	1.7976
20	0.035	1.4044	1.4231	1.5855	1.4980
21	0.032	1.2840	1.3011	1.4496	1.3696
22	0.028	1.1235	1.1385	1.2684	1.1984
23	0.025 or 1/40..........	1.0031	1.0165	1.1325	1.0700
24	0.022	0.8827	0.8945	0.9966	0.9416
25	0.02 or 1/50..........	0.8025	0.8132	0.9060	0.8560
26	0.018	0.7222	0.7319	0.8154	0.7704
27	0.016	0.6420	0.6506	0.7248	0.6848
28	0.014	0.5617	0.5692	0.6342	0.5992
29	0.013	0.5216	0.5286	0.5889	0.5564
30	0.012	0.4815	0.4879	0.5436	0.5136
31	0.01 or 1/100..........	0.4012	0.4066	0.4530	0.4280
32	0.009	0.3611	0.3659	0.4077	0.3852
33	0.008	0.3210	0.3253	0.3624	0.3424
34	0.007	0.2809	0.2846	0.3171	0.2996
35	0.005 or 1/200..........	0.2006	0.2033	0.2265	0.2140
36	0.004 or 1/250..........	0.1605	0.1626	0.1812	0.1712
	1.00 inch thick..........	41.5696	42.1236	46.9308	44.3408

TABLE NO. 74.

Weight of Square and Round Iron.

Side or Diam.	Weight, Square.	Weight, Round.	Side or Diam.	Weight, Square.	Weight, Round.	Side or Diam.	Weight, Square.	Weight, Round.
1/16	.013	.01	2	13.52	10.616	5	84.48	66.35
1/8	.053	.041	2 1/8	15.263	11.988	5 1/4	93.168	73.172
3/16	.118	.093	2 1/4	17.112	13.44	5 1/2	102.24	80.304
1/4	.211	.165	2 3/8	19.066	14.975	5 3/4	111.756	87.776
3/8	.475	.373	2 1/2	21.12	16.588	6	121.664	95.552
1/2	.845	.663	2 5/8	23.292	18.293	6 1/4	132.04	103.704
5/8	1.32	1.043	2 3/4	25.56	20.076	6 1/2	142.816	112.16
3/4	1.901	1.493	2 7/8	27.939	21.944	6 3/4	154.012	120.96
7/8	2.588	2.032	3	30.416	23.888	7	165.632	130.048
1	3.38	2.654	3 1/4	35.704	28.04	7 1/4	177.672	139.544
1 1/8	4.278	3.359	3 1/2	41.408	32.515	7 1/2	190.136	149.328
1 1/4	5.28	4.147	3 3/4	47.534	37.332	7 3/4	203.024	159.456
1 3/8	6.39	5.019		54.084	42.464	8	216.336	169.856
1 1/2	7.604	5.972	4	61.055	47.952	9	273.792	215.04
1 5/8	8.926	7.01	4 1/4	68.448	53.76			
1 3/4	10.352	8.128	4 3/4	76.264	59.9			
1 7/8	11.883	9.333						

TABLE NO. 75.

Vulgar Fractions of a Lineal Inch in Decimal Fractions.

	ADVANCING BY THIRTY-SECONDS.					ADVANCING BY ODD SIXTY-FOURTHS.			
Thirty-seconds.	Fractions.	Decimals of an Inch.	Thirty-seconds.	Fractions.	Decimals of an Inch.	Sixty-fourths.	Decimals of an Inch.	Sixty-fourths.	Decimals of an Inch.
1	1/32	0.03125	17	17/32	0.53125	1	0.015625	33	0.515625
2	1/16	0.0625	18	9/16	0.5625	3	0.04687	35	0.546875
3	3/32	0.09375	19	19/32	0.59375	5	0.078125	37	0.578125
4	1/8	0.125	20	5/8	0.625	7	0.109375	39	0.609375
5	5/32	0.15625	21	21/32	0.65625	9	0.140625	41	0.640625
6	3/16	0.1875	22	11/16	0.6875	11	0.171875	43	0.671875
7	7/32	0.21875	23	23/32	0.71875	13	0 203125	45	0.703125
8	1/4	0.25	24	3/4	0.75	15	0.234375	47	0.734375
9	9/32	0.28125	25	25/32	0.78125	17	0.265625	49	0.765625
10	5/16	0.3125	26	13/16	0.8125	19	0.296875	51	0.796875
11	11/32	0.34375	27	27/32	0.84375	21	0.328125	53	0.828125
12	3/8	0.375	28	7/8	0.875	23	0.359375	55	0.859375
13	13/32	0.40625	29	29/32	0.90625	25	0.390625	57	0.890625
14	7/16	0.4375	30	15/16	0.9375	27	0.421875	59	0.921875
15	15/32	0.46875	31	31/32	0.96875	29	0.453125	61	0.953125
16	1/2	0.5	32	1	1.000	31	0.484375	63	0.984375

Lineal Inches in Decimal Fractions of a Lineal Foot.

Lineal Inches.	Lineal Foot.	Lineal Inches.	Lineal Foot.	Lineal Inches.	Lineal Foot.
1/64	0.001302083	1 7/8	0.15625	6 1/2	0.5416
1/32	0.00260416	2	0.1666	6 3/4	0.5625
1/16	0.0052083	2 1/8	0.177083	7	0.5833
1/8	0.010416	2 1/4	0.1875	7 1/4	0.60416
3/16	0.015625	2 3/8	0.197916	7 1/2	0.625
1/4	0.02083	2 1/2	0.2083	7 3/4	0.64583
5/16	0.0260416	2 5/8	0.21875	8	0.66667
3/8	0.03125	2 3/4	0.22916	8 1/4	0.6875
7/16	0.0364583	2 7/8	0.239583	8 1/2	0.7083
1/2	0.0416	3	0.25	8 3/4	0.72916
9/16	0.046875	3 1/4	0.27083	9	0.75
5/8	0.052083	3 1/2	0.2916	9 1/4	0.77083
11/16	0.0572916	3 3/4	0.3125	9 1/2	0.7916
3/4	0.0625	4	0.33333	9 3/4	0.8125
13/16	0.0677083	4 1/4	0.35416	10	0.83333
7/8	0.072916	4 1/2	0.375	10 1/4	0.85416
15/16	0.078125	4 3/4	0.39583	10 1/2	0.875
1	0.0833	5	0.4166	10 3/4	0.89583
1 1/8	0.09375	5 1/4	0.4375	11	0.9166
1 1/4	0.10416	5 1/2	0.4583	11 1/4	0.9375
1 3/8	0.114583	5 3/4	0.47916	11 1/2	0.9583
1 1/2	0.125	6	0.5	11 3/4	0.97916
1 5/8	0.135416	6 1/4	0.52083	12	1.000
1 3/4	0.14583				

The *first cost* of a boiler is a fixed quantity. The *cost of operation* is one continuing during the life of the boiler. Given the relative cost of tubular and water-tube boilers, and the cost of fuel, it is a simple arithmetical calculation to determine what percentage of economy there must be in water-tube boilers in order to earn back their extra first cost. Of course no one who understands the subject, now doubts that there is *some* advantage in water-tube boilers in point of economy of operation and repairs. Take this percentage of economy at the minimum—say only 10%—and see how short a time it takes to amount to more than the cost of the boiler. It will surprise you.

Square Inches in Decimal Fractions of a Square Foot.

Square Inches.	Square Foot.	Square Inches.	Square Foot.	Square Inches.	Square Foot.	Square Inches.	Square Foot.
0.10	0.0006944	24.0	0.16666	65.0	0.45138	105.0	0.72916
0.15	0.0010416	25.0	0.17361	66.0	0.45833	106.0	0.73611
0.20	0.001388	26.0	0.18055	67.0	0.46527	107.0	0.74305
0.25	0.0017361	27.0	0.18750	68.0	0.47222	108.0	0.75000
0.30	0.002083	28.0	0.19444	69.0	0.47916	109.0	0.75694
0.35	0.0024305	29.0	0.20138	70.0	0.48611	110.0	0.76388
0.40	0.002777	30.0	0.20833	71.0	0.49305	111.0	0.77083
0.45	0.00311249	31.0	0.21527	72.0	0.50000	112.0	0.77777
0.50	0.003472	32.0	0.22222	73.0	0.50694	113.0	0.78472
0.55	0.0038194	33.0	0.22916	74.0	0.51388	114.0	0.79166
0.60	0.004166	34.0	0.23611	75.0	0.52083	115.0	0.79861
0.65	0.0045138	35.0	0.24305	76.0	0.52777	116.0	0.80555
0.70	0.004861	36.0	0.25000	77.0	0.53472	117.0	0.81249
0.75	0.0052083	37.0	0.25694	78.0	0.54166	118.0	0.81944
0.80	0.005555	38.0	0.26388	79.0	0.54861	119.0	0.82638
0.85	0.0059027	39.0	0.27083	80.0	0.55555	120.0	0.83333
0.90	0.006250	40.0	0.27777	81.0	0.56249	121.0	0.84027
0.95	0.0065972	41.0	0.28472	82.0	0.56944	122.0	0 84722
1.0	0.006944	42.0	0.29166	83.0	0.57638	123.0	0.85416
2.0	0.01388	43.0	0.29861	84.0	0.58333	124.0	0.86111
3.0	0.02083	44.0	0.30555	85.0	0.59027	125.0	0.86805
4.0	0.02777	45.0	0.31249	86.0	0.59722	126.0	0.87500
5.0	0.03472	46.0	0.31944	87.0	0.60416	127.0	0.88194
6.0	0.04166	47.0	0.32638	88.0	0.61111	128.0	0.88888
7.0	0.04861	48.0	0.33333	89.0	0.61805	129.0	0.89583
8.0	0.05555	49.0	0.34027	90.0	0.62500	130.0	0.90277
9.0	0.06250	50.0	0.34722	91.0	0.63194	131.0	0.90972
10.0	0.06944	51.0	0.35416	92.0	0.63888	132.0	0.91666
11.0	0.07638	52.0	0.36111	93.0	0.64583	133.0	0.92361
12.0	0.08333	53.0	0.36805	94.0	0.65277	134.0	0.93055
13.0	0.09027	54.0	0.37500	95.0	0.65972	135.0	0.93750
14.0	0.09722	55.0	0.38194	96.0	0.66666	136.0	0.94444
15.0	0.10416	56.0	0.38888	97.0	0.67361	137.0	0.95138
16.0	0.11111	57.0	0.39583	98.0	0.68055	138.0	0.95833
17.0	0.11805	58.0	0.40277	99.0	0.68750	139.0	0.96527
18.0	0.12500	59.0	0.40972	100.0	0.69444	140.0	0.97222
19.0	0.13194	60.0	0.41666	101.0	0.70138	141.0	0.97916
20.0	0.13888	61.0	0.42361	102.0	0.70833	142.0	0.98611
21.0	0.14583	62.0	0.43055	103.0	0.71527	143.0	0.99305
22.0	0.15277	63.0	0.43750	104.0	0.72222	144.0	1.0000
23.0	0.15972	64.0	0.44444				

Decimal Fractions of a Square Foot in Square Inches.

Square Foot.	Square Inches.	Square Foot.	Square Inches.	Square Foot.	Square Inches.	Square Foot.	Square Inches.
0.01	1.44	0.26	37.4	0.51	73.4	0.76	109.4
0.02	2.88	0.27	38.9	0.52	74.9	0.77	110.9
0.03	4.32	0.28	40.3	0.53	76.3	0.78	112.3
0.04	5.76	0.29	41.8	0.54	77.8	0.79	113.8
0.05	7.20	0.30	43.2	0.55	79.2	0.80	115.2
0.06	8.64	0.31	44.6	0.56	80.6	0.81	116.6
0.07	10.1	0.32	46.1	0.57	82.1	0.82	118.1
0.08	11.5	0.33	47.5	0.58	83.5	0.83	119.5
0.09	13.0	0.34	49.0	0.59	85.0	0.84	121.0
0.10	14.4	0.35	50.4	0.60	86.4	0.85	122.4
0.11	15.8	0.36	51.8	0.61	87.8	0.86	123.8
0.12	17.3	0.37	53.3	0.62	89.3	0.87	125.3
0.13	18.7	0.38	54.7	0.63	90.7	0.88	126.7
0.14	20.2	0.39	56.2	0.64	92.2	0.89	128.2
0.15	21.6	0.40	57.6	0.65	93.6	0.90	129.6
0.16	23.0	0.41	58.0	0.66	95.0	0.91	131.0
0.17	24.5	0.42	60.5	0.67	96.5	0.92	132.5
0.18	25.9	0.43	61.9	0.68	97.9	0.93	133.9
0.19	27.4	0.44	63.4	0.69	99.4	0.94	135.4
0.20	28.8	0.45	64.8	0.70	100.8	0.95	136.8
0.21	30.2	0.46	66.2	0.71	102.2	0.96	138.2
0.22	31.7	0.47	67.7	0.72	103.7	0.97	139.7
0.23	33.1	0.48	69.1	0.73	105.1	0.98	141.1
0.24	34.6	0.49	70.6	0.74	106.6	0.99	142.6
0.25	36.0	0.50	72.0	0.75	108.0	1.00	144.0

How many large modern boiler plants are now constructed with old style flue and tubular boilers—boilers in which circulation is in spite of, and not because of, their design and construction? Among the big new installations there are twenty water-tube plants now to every one of the old style. Yet many small boiler users still fail to grasp the fact that the economy of water-tube boilers is "a condition" and not "a theory."

French Measures of Length with U. S. Equivalents.

		Metres.	U. S. Equivalents.
	1 millimetre	0.001	0.03937 inch.
10 millimetres	1 centimetre	0.01	0.3937 inch.
10 centimetres	1 decimetre	0.1	3.93704 inches.
10 decimetres 100 centimetres 1000 millimetres	} 1 METRE	1.0	{ 39.3704 inches. 3.2809 feet.
10 metres	1 decametre	10.0	32.8087 feet.
10 decametres	1 hectometre	100.0	328.0869 feet.
10 hectometres	1 KILOMETRE	1000.0	3280.869 feet.
10 kilometres	1 myriametre	10000.0	6.21377 miles.

French Measures of Surface with U. S. Equivalents.

		Square Metres	U. S. Equivalents.
	1 sq. millimetre	0.000001	0.00155 sq. inches.
100 sq. millimetres	1 sq. centimeter	0.0001	0.155 sq. inches.
100 sq. centimetres	1 sq. decimetre	0.01	15.5003 sq. inches.
100 sq. decimetres 10000 sq. centimetres	} 1 sq. METRE	1.0	{ 10.7641 sq. feet. 1.1960 sq. yards.
100 sq. metres	1 sq. decametre	100.0	{ 1076.41 sq. feet. 119.601 sq. yards.
100 sq. decametres	1 sq. hectometre	10,000.0	{ 11960.11 sq. yards. 2.4711 acres.
100 sq. hectometres	1 sq. kilometre	1,000,000.0	{ 1196014 sq. yards. 0.38611 sq. miles.
100 sq. kilometres	1 sq. myriametre	100,000,000.0	38.611 sq. miles.

French Measures of Weight with U. S. Avoirdupois Equivalents.

		Grammes.	U. S. Equivalents.
	1 milligramme	0.001	0.0154 grains.
10 milligrammes	1 centigramme	0.01	0.1543 grains.
10 centigrammes	1 decigramme	0.1	1.5432 grains.
10 decigrammes	1 GRAMME	1.0	15.4323 grains.
10 grammes	1 decagramme	10.0	{ 154.3235 grains. 0.3527 ounces.
10 decagrammes	1 hectagramme	100.0	{ 1543.2349 grains. 3.5274 ounces.
10 hectagrammes	1 kilogramme	1000.0	2.2046 pounds.
10 kilogrammes	1 metric quintal		220.4621 pounds.
10 quintals 1000 kilogrammes	} 1 millier or tonne		{ 2204.6212 pounds. 19.6841 cwt. 0.9842 tons.

French Measures of Volume with U. S. Equivalents.

		Cubic Metres.	U. S Equivalents.
	1 cu. millimetre	0.000000001	0.000061 cu. inches.
1000 cu. millimetres	1 cu. centimetre	0.000001	0.061025 cu. inches.
1000 cu. centimetres	1 cu. decimetre	0.001	61.02524 cu. inches. / 0.0353156 cu. feet.
1000 cu. decimetres	1 cu. METRE	1.0	35.3156 cu. feet. / 1.308 cu. yards.
1000 cu. metres	1 cu. decametre	1000	1308.0 cu. yards.

French Liquid Measure with U. S. Equivalents.

		Litres.	U. S. Equivalents.
	1 centilitre / 10 cu. centimetres	0.01	0.61025 cu. inches. / 0.0845 gills.
10 centilitres	1 decilitre	0.1	6.1025 cu. inches. / 0.2114 pints.
10 decilitres	1 LITRE / 1 cu. decimetre	1.0	61.02524 cu. inches. / 0.2642 gallons.
10 litres	1 decalitre	10.0	2.6418 gallons.
10 decalitres	1 hectolitre	100.0	26.418 gallons.

THE FAMOUS SCHICHAU ENGINE.
Now owned by the C. C. Washburn Flouring Mills Co.
Steam supplied by Heine Boilers.

TABLE No. 84.

Wrought Iron Welded Steam, Gas and Water Pipe.—Table of Standard Dimensions.

Diameter			Thickness.	Circumference.		Transverse Areas.			Length of Pipe per Square Foot of		Length of Pipe Containing One Cubic Foot.	Nominal Weight per Foot.	Number of Threads per Inch of Screw.
Nominal Internal.	Actual External.	Actual Internal.		External.	Internal.	External.	Internal.	Metal.	External Surface.	Internal Surface.			
Inches.	Inches.	Inches.	Inches.	Inches.	Inches.	Sq. Inches.	Sq. Inches.	Sq. Inches.	Feet.	Feet.	Feet.	Pounds.	
1/8	.405	.27	.068	1.272	.848	.129	.0573	.0717	9.44	14.15	2513.	.241	27
1/4	.54	.364	.088	1.696	1.144	.229	.1041	.1249	7.075	10.49	1383.3	.42	18
3/8	.675	.494	.091	2.121	1.552	.358	.1917	.1663	5.657	7.73	751.2	.559	18
1/2	.84	.623	.109	2.639	1.957	.554	.3048	.2492	4.547	6.13	472.4	.837	14
3/4	1.05	.824	.113	3.299	2.589	.866	.5333	.3327	3.637	4.635	270.	1.115	14
1	1.315	1.048	.134	4.131	3.292	1.358	.8626	.4954	2.904	3.645	166.9	1.608	11½
1¼	1.66	1.38	.14	5.215	4.335	2.164	1.496	.668	2.301	2.768	96.25	2.244	11½
1½	1.9	1.611	.145	5.969	5.061	2.835	2.038	.797	2.01	2.371	70.66	2.678	11½
2	2.375	2.067	.154	7.461	6.494	4.43	3.356	1.074	1.608	1.848	42.91	3.609	11½
2½	2.875	2.468	.204	9.032	7.753	6.492	4.784	1.708	1.328	1.547	30.1	5.739	11½
3	3.5	3.067	.217	10.996	9.636	9.621	7.388	2.243	1.091	1.245	19.5	7.536	8
3½	4.	3.548	.226	12.566	11.146	12.566	9.887	2.679	.955	1.077	14.57	9.001	8
4	4.5	4.026	.237	14.137	12.648	15.904	12.73	3.174	.849	.949	11.31	10.665	8
4½	5.	4.508	.246	15.708	14.162	19.635	15.961	3.674	.764	.848	9.02	12.34	8
5	5.563	5.045	.259	17.477	15.849	24.306	19.99	4.316	.687	.757	7.2	14.502	8
6	6.625	6.065	.28	20.813	19.054	34.472	28.888	5.584	.577	.63	4.98	18.762	8
7	7.625	7.023	.301	23.955	22.063	45.664	38.738	6.926	.501	.544	3.72	23.271	8
8	8.625	7.982	.322	27.046	25.076	58.426	50.04	8.386	.443	.478	2.88	28.177	8
9	9.625	8.937	.344	30.238	28.076	72.76	62.73	10.03	.397	.427	2.29	33.701	8
10	10.75	10.019	.366	33.772	31.477	90.763	78.859	11.924	.355	.382	1.82	40.065	8
11	12.	11.25	.375	37.699	35.343	113.098	99.402	13.696	.318	.339	1.456	45.95	8
12	12.75	12.	.375	40.065	37.7	127.677	113.098	14.579	.299	.319	1.27	48.985	8
13	14.	13.25	.375	43.982	41.626	153.938	137.887	16.051	.273	.288	1.04	53.921	8
14	15.	14.25	.375	47.124	44.768	176.715	159.485	17.23	.255	.268	.903	57.893	8
15	16.	15.25	.375	50.265	47.909	201.062	182.655	18.407	.239	.250	.788	61.77	8

Nominal Internal	Actual External	Actual Internal	Thickness	Circumference External	Circumference Internal	Transverse Areas External	Transverse Areas Internal	Transverse Areas Metal	Length ext. surface	Length int. surface	Length of Pipe Containing One Cubic Foot	Nominal Weight per Foot	Number of Threads per Inch of Screw
Inches.	Inches.	Inches.	Inches.	Inches.	Inches.	Sq. Inches.	Sq. Inches.	Sq. Inches.	Feet.	Feet.	Feet.	Pounds.	No.
--	18.	17.25	.375	56.549	54.192	254.47	233.706	20.764	.212	.221	.616	69.66	--
--	20.	19.25	.375	62.832	60.476	314.16	291.04	23.12	.191	.198	.495	77.57	--
--	22.	21.25	.375	69.115	66.759	380.134	354.657	25.477	.174	.179	.406	85.47	--
--	24.	23.25	.375	75.398	73.042	452.39	424.558	27.832	.159	.164	.339	93.37	--

TABLE NO. 85.

Wrought Iron Welded Extra Strong Pipe.—Table of Standard Dimensions.

Nominal Internal	Actual External	Actual Internal	Thickness	Nearest Wire Gauge	Circumference External	Circumference Internal	Transverse Areas External	Transverse Areas Internal	Transverse Areas Metal	Length External Surface	Length Internal Surface	Nominal Weight per Foot
Inches.	Inches.	Inches.	Inches.	No.	Inches.	Inches.	Sq. Inches.	Sq. Inches.	Sq. Inches.	Feet.	Feet.	Pounds.
1/8	.405	.205	.1	12½	1.272	.644	.129	.033	.086	9.433	18.632	.29
1/4	.54	.294	.123	11	1.696	.924	.229	.068	.161	7.075	12.986	.54
3/8	.675	.421	.127	10½	2.121	1.323	.358	.139	.219	5.657	9.07	.74
1/2	.84	.542	.149	9	2.639	1.703	.554	.231	.323	4.547	7.046	1.09
3/4	1.05	.736	.157	8½	3.299	2.312	.866	.452	.414	3.637	5.109	1.39
1	1.315	.951	.182	7	4.131	2.988	1.358	.71	.648	2.904	4.016	2.17
1 1/4	1.66	1.272	.194	6½	5.215	3.996	2.164	1.271	.893	2.301	3.003	3.
1 1/2	1.9	1.494	.203	6	5.969	4.694	2.835	1.753	1.082	2.01	2.556	3.63
2	2.375	1.933	.221	5	7.461	6.073	4.43	2.935	1.495	1.608	1.975	5.02
2 1/2	2.875	2.315	.28	2	9.032	7.273	6.492	4.209	2.283	1.328	1.649	7.07
3	3.5	2.892	.304	1	10.996	9.085	9.621	6.569	3.052	1.091	1.328	10.25
3 1/2	4.	3.358	.321	0	12.566	10.549	12.566	8.856	3.71	.955	1.137	12.47
4	4.5	3.818	.341	0	14.137	11.995	15.904	11.449	4.455	.849	1.	14.97
5	5.563	4.813	.375	00	17.477	15.120	24.306	18.193	6.12	.687	.793	20.54
6	6.625	5.75	.437	000	20.813	18.064	34.472	25.967	8.505	.577	.664	28.58

TABLE NO. 86.

Wrought Iron Welded Double Extra Strong Pipe.—Table of Standard Dimensions.

Diameter. Nominal Internal.	Diameter. Actual External.	Diameter. Actual Internal.	Thickness.	Nearest Wire Gauge.	Circumference. External.	Circumference. Internal.	Transverse Areas. External.	Transverse Areas. Internal.	Transverse Areas. Metal.	Length of Pipe per Square Foot of External Surface.	Length of Pipe per Square Foot of Internal Surface.	Nominal Weight per Foot.
Inches.	Inches.	Inches.	Inches.	No.	Inches.	Inches.	Sq. Inches.	Sq. Inches.	Sq. Inches.	Feet.	Feet.	Pounds.
1/2	.84	.244	.298	1	2.639	.766	.554	.047	.507	4.547	15.667	1.7
3/4	1.05	.422	.314	1	3.299	1.326	.866	.139	.727	3.637	9.049	2.44
1	1.315	.587	.364	00	4.131	1.844	1.358	.271	1.087	2.904	6.508	3.65
1¼	1.66	.885	.388	00	5.215	2.78	2.164	.615	1.549	2.304	4.317	5.2
1½	1.9	1.088	.406	000	5.969	3.418	2.835	.98	1.905	2.01	3.511	6.4
2	2.375	1.491	.442	0000	7.461	4.684	4.43	1.744	2.686	1.608	2.561	9.02
2½	2.875	1.755	.560	9/16 —	9.032	5.513	6.492	2.419	4.073	1.328	2.176	13.68
3	3.5	2.284	.608	5/8 —	10.996	7.175	9.621	4.097	5.524	1.091	1.672	18.56
3½	4.	2.716	.642	5/8 +	12.566	8.533	12.566	5.794	6.772	.955	1.406	22.75
4	4.5	3.136	.682	11/16 —	14.137	9.852	15.904	7.724	8.18	.849	1.217	27.48
5	5.563	4.063	.75	3/4	17.477	12.764	24.306	12.965	11.34	.687	.940	38.12
6	6.625	4.875	.875	7/8	20.813	15.315	34.472	18.666	15.806	.577	.784	53.11

TABLE NO. 87.

Lap-Welded Charcoal Iron Boiler Tubes.—Table of Standard Dimensions.

Diameter. External.	Diameter. Internal.	Thickness.	Wire Gauge.	Circumference. External.	Circumference. Internal.	Transverse Areas. External.	Transverse Areas. Internal.	Transverse Areas. Metal.	Length of Tube per Square Foot of External Surface.	Length of Tube per Square Foot of Internal Surface.	Nominal Weight per Foot.
Inches.	Inches.	Inches.	No.	Inches.	Inches.	Sq. Inches.	Sq. Inches.	Sq. Inches.	Feet.	Feet.	Pounds.
3	2.782	.109	12	9.425	8.74	7.069	6.079	.99	1.273	1.373	3.33
3¼	3.01	.12	11	10.21	9.456	8.296	7.116	1.18	1.175	1.26	3.96
3½	3.26	.12	11	10.996	10.241	9.621	8.347	1.274	1.091	1.172	4.28
3¾	3.51	.12	11	11.781	11.027	11.045	9.676	1.369	1.018	1.088	4.6
4	3.732	.134	10	12.566	11.724	12.566	10.939	1.627	.955	1.024	5.47
4¼	3.982	.134	10	13.352	12.51	14.186	12.453	1.733	.899	.959	5.82
4½	4.232	.134	10	14.137	13.295	15.904	14.066	1.838	.849	.902	6.17
4¾	4.482	.134	10	14.923	14.081	17.721	15.777	1.944	.804	.852	6.53
5	4.704	.148	9	15.708	14.778	19.635	17.379	2.256	.764	.812	7.58
5¼	4.954	.148	9	16.493	15.563	21.648	19.275	2.373	.728	.771	7.97
5½	5.204	.148	9	17.279	16.349	23.758	21.27	2.488	.694	.734	8.36
6	5.67	.165	8	18.85	17.813	28.274	25.249	3.025	.637	.673	10.16

Bowling Green Office Building,
NEW YORK.
Equipped with 750 H. P. Heine Safety Boilers.

LIST OF HEINE BOILER PLANTS.

HOTELS, OFFICES AND PUBLIC BUILDINGS.

National Guard Armory, St. Louis	1	Boiler,	40 H.P.
Harmonie Club, St. Louis	1	"	15 "
Shaw Building, St. Louis	1	"	50 "
Mitchell Building, St. Louis	1	"	45 "
Missouri State University, Columbia, Mo	2	"	120 "
St. Louis Exposition, St. Louis	4	"	1000 "
House of Refuge, St. Louis	2	"	110 "
Montesano Hotel, St. Louis	1	"	15 "
Palace Hotel, San Francisco, Cal	4	"	260 "
McVicker's Theatre, Chicago, Ill	2	"	260 "
Roe Building, St. Louis	2	"	250 "
Good Samaritan Hospital, St. Louis	1	"	20 "
Lake View School House, Chicago, Ill	2	"	75 "
State University, Madison, Wis	2	"	110 "
Minneapolis Industrial Exposition, Minneapolis, Minn	4	"	1000 "
Centropolis Hotel, Kansas City	2	"	150 "
Inter-State Industrial Exposition, Chicago, Ill	1	"	250 "
University of Denver, Denver, Colo., first order	1	"	90 "
University of Denver, Denver, Colo., second order	1	"	200 "
Boston Heating Co., Boston, Mass., first order	10	"	1000 "
Boston Heating Co., Boston, Mass., second order	2	"	200 "
San Jose Insane Asylum, San Jose, Cal	2	"	220 "
Cincinnati Exposition, Cincinnati, Ohio	2	"	400 "
Denver Republican, Denver, Colo	1	"	80 "
Cathedral of St. John, Denver, Colo	1	"	80 "
Railroad Building, Denver, Colo	1	"	110 "
Arkansas State Lunatic Asylum, Little Rock, Ark	2	"	150 "
Western Pennsylvania Exposition, Pittsburg, Pa	2	"	500 "
Houser Building, St. Louis	2	"	240 "
Hospital for the Insane, Fergus Falls, Minn	1	"	120 "
Southern Illinois Penitentiary, Chester, Ill	2	"	600 "
Insane Asylum, Agnew, Cal	2	"	220 "
Railroad Building, Denver, Colo., second order	1	"	110 "
Court House, Evansville, Ind	2	"	190 "
Wm. Kirkup & Son, Cincinnati, Ohio	2	"	220 "
Pabst Opera House, Milwaukee, Wis	2	"	240 "
Arapahoe County Jail, Denver, Colo	3	"	360 "
C. D. McPhee, Denver, Colo	2	"	150 "
Broadway Theatre, Denver, Colo., first order	2	"	240 "
Boatmen's Bank Building, St. Louis	2	"	240 "
Broadway Theatre, Denver, Colo., second order	1	"	120 "
Samuel Cupples Real Estate Co., St. Louis	2	"	740 "
The Neave Building Co., Cincinnati, Ohio	2	"	240 "
First National Bank, Duluth, Minn	2	"	80 "
John Shillito Co., Cincinnati, Ohio, first order	1	"	250 "
John Shillito Co., Cincinnati, Ohio, second order	2	"	400 "
John M. Smyth, Chicago, Ill	2	"	500 "
University of Michigan, Ann Arbor, Mich	2	"	300 "
Hospital for the Insane, Fergus Falls, Minn., second order	1	"	120 "
Palace Hotel, San Francisco, Cal., second order	3	"	525 "
Royal Victoria Hospital, Montreal, Can	2	"	160 "
R. H. Macy & Co., New York	3	"	600 "
Henry C. Brown, Palace Hotel, Denver, Colo	4	"	520 "
Jas. G. Fair Building, San Francisco, Cal	1	"	175 "
Hospital for the Insane, Fergus Falls, Minn., third order	1	"	120 "
Jackson County Court House, Kansas City	3	"	360 "
Equitable Building, Des Moines, Iowa	2	"	300 "
Freehold Building, Toronto, Can	1	"	120 "
Confederation Building, Toronto, Can	2	"	200 "
John Doty Engine Co., Toronto, Can	1	"	110 "
Athletic Club Building, Chicago, Ill	2	"	300 "
World's Columbian Exposition, Chicago, Ill	8	"	3000 "
Windsor Hotel, Montreal, Can	1	"	150 "
Betz Building, Philadelphia, Pa	3	"	500 "
Mercantile Club, St. Louis	2	"	300 "
The Johnson Building, Cincinnati, Ohio	2	"	240 "
The First National Bank, Pittsburg, Pa	2	"	100 "
Jackson County Court House, Kansas City, second order	1	"	120 "
Emma Spreckels Building, San Francisco, Cal	2	"	150 "
Mallinckrodt Building, St. Louis	2	"	300 "

World's Columbian Exposition, Chicago, Ill., second order	4 Boiler,	1500 H.P.
Odd Fellows' Temple, Cincinnati, Ohio	3 "	450 "
Occidental Hotel, San Francisco, Cal	2 "	150 "
New Planters' House, St. Louis	2 "	600 "
Young Men's Christian Association Building, Chicago	2 "	500 "
Gruenewald Building, New Orleans	2 "	100 "
Hospital for the Insane, Fergus Falls, Minn., fourth order	1 "	120 "
Pennsylvania State College, State College, Pa	1 "	150 "
Hotel Majestic, New York City	4 "	1000 "
Carnegie Steel Co., Pittsburg, Pa., first order	2 "	250 "
Carnegie Steel Co., Pittsburg, Pa., second order	1 "	255 "
Central Park Apartment Building, New York City	2 "	500 "
California Midwinter International Exposition, San Francisco	8 "	3000 "
City and County Building, Salt Lake City, Utah	3 "	225 "
Southern Illinois Penitentiary, Chester, Ill., second order	1 "	300 "
La Banque du Peuple, Montreal, Can	2 "	125 "
St. Vincent's Institute, Normandy, Mo	2 "	400 "
Marquette Building, Chicago, Ill	4 "	1000 "
Merchants' Exchange, St. Louis	3 "	360 "
Fergus Falls State Hospital, Fergus Falls, Minn., fifth order	1 "	120 "
City of Chicago Electric Light Station, Chicago, Ill	3 "	1500 "
Bennett & Wright, Parliament Building, Victoria, B. C	2 "	130 "
New Planters' House, St. Louis, second order	1 "	200 "
Equitable Building, Denver, Colo	3 "	750 "
Samuel Cupples Real Estate Co., St. Louis, second order	1 "	150 "
Jackson Bros. Building, Pittsburg, Pa	2 "	200 "
Maine State College, Orona, Me	1 "	85 "
Cornell University, Ithaca, N. Y	1 "	100 "
First National Bank Building, Chicago, Ill	2 "	408 "
North Sub-district School, Pittsburg, Pa	2 "	130 "
University of Missouri, Columbia, Mo	1 "	200 "
Winnebago Building, Chicago, Ill	3 "	375 "
Bowling Green Building, New York	3 "	540 "
Central Kentucky Lunatic Asylum, Lakeland, Ky	1 "	300 "
Y. M. C. A. Building, St. Louis	1 "	120 "
First National Bank Building, Pittsburg, Pa	1 "	150 "
Colorado Telephone Co., Denver, Colo	1 "	60 "
W. U. Telegraph Co. Building, Chicago, Ill	1 "	250 "
Lindell Real Estate Co., St. Louis	3 "	450 "
Fergus Falls State Hospital, Fergus Falls, Minn., sixth order	1 "	120 "
New City Hall, St. Louis	2 "	500 "
Uihlein Building, Milwaukee, Wis	2 "	360 "
Windsor Hotel, New York	2 "	356 "
C. S. Movey Mercantile Co., Denver, Colo	2 "	120 "
Y. M. C. A., Philadelphia, Pa	2 "	154 "
Massillon State Hospital, Massillon, Ohio	4 "	1000 "
Davis Building, St. Louis, Mo	4 "	600 "
Hotel Chamberlain, Old Point Comfort, Va	1 "	302 "
Quincy House, Boston, Mass	1 "	210 "
Fullerton Building, St. Louis, Mo	2 "	300 "
Arkansas State Lunatic Asylum, Little Rock, Ark., second order,	1 "	75 "
California Hotel, San Francisco	1 "	140 "
Hospital St. Jean de Dieu Insane Asylum, Quebec	1 "	80 "
Municipal and County Buildings, Toronto, Ont	4 "	700 "
Forresters' Temple, Toronto	2 "	240 "

IRON AND STEEL MANUFACTURERS.

St. Louis Stamping Co., St. Louis	1 Boiler,	40 H.P.
Troy Steel and Iron Co., Troy, N. Y., first order	2 "	175 "
Troy Steel and Iron Co., Troy, N. Y., second order	8 "	2560 "
Risdon Iron Works, San Francisco, Cal	3 "	600 "
Tudor Iron Works, St. Louis	2 "	500 "
Scherpe & Koken, St. Louis	1 "	30 "
Shoenberger & Co., Pittsburg, Pa	1 "	250 "
Oliver Bros. & Phillips, Pittsburg, Pa., first order	2 "	500 "
Edgar Thompson Steel Works, Braddock, Pa	2 "	500 "
Union Steel Co., Chicago, Ill	2 "	500 "
Jas. P. Witherow, Pittsburg, Pa	2 "	250 "
Troy Steel and Iron Co., Troy, N. Y., third order	1 "	125 "
Belleville Nail Co., Belleville, Ill	5 "	1250 "
Racine Hardware Co., Racine, Wis	1 "	150 "
Robt. Hare Powel's Sons, Saxton, Pa	3 "	750 "
Valentine Ore Land Association, Bellefonte, Pa	3 "	375 "
Troy Steel and Iron Co., Troy, N. Y., fourth order	2 "	180 "

Chicago Steel Works, Chicago, Ill	1 Boiler,	150 H.P.
Troy Steel and Iron Co., Troy, N. Y., fifth order	3 "	375 "
North Branch Steel Co., Danville, Pa	4 "	1280 "
Oliver Bros. & Phillips, Pittsburg, Pa., second order	1 "	250 "
Henry Disston & Sons, Philadelphia, Pa., first order	1 "	200 "
Springfield Iron Co., Springfield, Ill	3 "	750 "
H. R. Worthington, Brooklyn, N. Y.	2 "	300 "
Henry Disston & Sons, Philadelphia, Pa., second order	3 "	750 "
Oliver Bros. & Phillips, Pittsburg, Pa., third order	1 "	250 "
Missouri Malleable Iron Co., St. Louis	1 "	150 "
Helmbacher Forge and Rolling Mill Co., St. Louis	1 "	250 "
Chester Rolling Mills, Chester, Pa	5 "	1250 "
Van Zile, McCormack & Co., Albany, N. Y.	1 "	150 "
Troy Steel and Iron Co., Troy, N. Y., sixth order	2 "	640 "
Muskegon Iron and Steel Co., Muskegon, Mich	1 "	200 "
Muskegon Iron and Steel Co., Muskegon, Mich., second order,	1 "	250 "
Monterey Foundry Co., Monterey, Mex	1 "	65 "
Van Zile, McCormack & Co., Albany, N. Y., second order	1 "	150 "
Kilmer Mfg. Co., Newburgh, N. Y	5 "	750 "
The Johnson Co., Johnstown, Pa	2 "	500 "
Strom Mfg. Co., Chicago, Ill	1 "	120 "
Keystone Rolling Mill Co., Pittsburg, Pa	1 "	150 "
St. Louis Shovel Co., St. Louis	1 "	200 "
U. S. Iron and Tin Plate Mfg. Co., Demmler, Pa	1 "	250 "
Shoenberger, Speer & Co., Pittsburg, Pa	1 "	300 "
Missouri Malleable Iron Co., St. Louis, second order	2 "	300 "
Elba Iron Works, Pittsburg, Pa	1 "	150 "
Keystone Rolling Mill Co., Pittsburg, Pa., second order	1 "	375 "
Scherpe & Koken Arch. Iron Works Co., St. Louis, 2d order	2 "	120 "
The Johnson Co., Johnstown, Pa., second order	2 "	500 "
Addyston Pipe and Steel Co., Addyston, Ohio	1 "	300 "
Helmbacher Forge & Rolling Mill Co., St. Louis, second order,	1 "	375 "
U. S. Iron and Tin Plate Mfg. Co., Demmler, Pa., second order,	1 "	375 "
Tudor Iron Works, St. Louis, second order	2 "	500 "
Illinois Steel Co., Joliet, Ill	4 "	1000 "
U. S. Iron and Tin Plate Mfg. Co., Demmler, Pa., third order.	4 "	400 "
Illinois Steel Co., Chicago, Ill., second order	4 "	1000 "
S. T. Williams & Co., Muscatine, Iowa	1 "	300 "
Jas. McKinney & Son, Albany, N. Y	1 "	60 "
Williams Rolling Mills, Muscatine, Iowa, second order	1 "	120 "
Tudor Iron Works, East St. Louis, third order	2 "	750 "
Session's Foundry Co., Bristol, Conn	3 "	405 "
Illinois Steel Co., Joliet, Ill., third order	2 "	500 "
Illinois Steel Co., Joliet, Ill., fourth order	4 "	1000 "
Jones & Laughlin, L't'd, Pittsburg, Pa	1 "	500 "
Schoenberger Steel Co., Pittsburg, Pa., second order	1 "	300 "
Koken Iron Works, St. Louis, second order	1 "	175 "
Inland Steel Co., Chicago Heights, Ill	1 "	150 "

ARTIFICIAL ICE AND REFRIGERATOR COMPANIES.

Texarkana Ice Co., Texarkana, Ark	2 Boiler,	150 H.P.
Bohlen-Huse Machine and Lake Ice Co., Memphis, Tenn	1 "	110 "
Griesedieck Artificial Ice Co., St. Louis, first order	1 "	250 "
H. H Bodeman, St. Louis	1 "	120 "
Griesedieck Artificial Ice Co., St. Louis, second order	1 "	250 "
Springfield Ice and Refrigerator Co., Springfield, Mo	1 "	120 "
H. Henke & Co., Houston, Tex	1 "	120 "
Union Ice Mfg. Co., Pittsburg, Pa	2 "	600 "
New York Hygeia Ice Co., New York, N. Y	2 "	750 "
St. Joseph Artificial I. and C S. Co., St. Joseph, Mo	1 "	150 "
Union Ice Mfg. Co., Pittsburg, Pa., second order	1 "	300 "
H. Henke & Co., Houston, Tex., second order	1 "	300 "
New York Hygeia Ice Co., New York, N. Y., second order	1 "	375 "
H. Henke Artificial Ice Co., Houston, Tex., third order	1 "	300 "

FLOUR MILLS.

Del Monte Flour Mill, San Francisco, Cal	1 Boiler,	180 H.P.
Chas. Tiedemann, Collinsville, Ill	1 "	125 "
Little Rock Milling and Elevator Co., Little Rock, Ark	1 "	170 "
Parson's Flour Mills, San Francisco, Cal	1 "	145 "
Capitol Flour Mills, Los Angeles, Cal., first order	1 "	140 "
Texas Star Flour Mills, Galveston, Tex	1 "	200 "
J. B. Thro & Co., St. Charles, Mo	1 "	110 "
Eisenmayer Milling and Elevator Co., Halstead, Kan	1 "	200 "
Hardesty Bros., Columbus, Ohio	1 "	200 "

L. Hoster Brewing Co.,
COLUMBUS, O.
Equipped with 2320 H. P. Heine Boilers.

Lyon, Clement & Greenleaf, Ligonier, Ind	1 Boiler,	150	H.P.
Koppitz & Smith, Pacific, Mo	1 "	60	"
Capitol Milling Co., Los Angeles, Cal., second order	1 "	155	"
Colorado Milling and Elevator Co., Denver, Colo	1 "	250	"
The H. C. Cole Milling Co., Chester, Ill	1 "	300	"
G. Ziebold & Son, Red Bud, Ill	1 "	150	"
Glenn Bros., Hillsboro, Ill	1 "	120	"
Boney & Harper, Wilmington, N. C	1 "	120	"
Blish Milling Co., Seymour, Ind	1 "	225	"
Franz Huning, Glorietta Mills, Albuquerque, N. M	1 "	80	"
Taylor Bros. & Co., Quincy, Ill	2 "	500	"
Farmers' Union and Milling Co., Stockton, Cal	2 "	500	"
R. T. Davis Mill Co., St. Joseph, Mo	1 "	250	"
The Cerealine Mfg. Co., Indianapolis, Ind	1 "	375	"
The Cerealine Mfg. Co., Indianapolis, Ind., second order	1 "	250	"
Plymouth Roller Mill Co., LeMars, Iowa	1 "	250	"
The Cerealine Mfg. Co., Indianapolis, Ind., third order	1 "	375	"
McDaniel & Co., Franklin, Ind	1 "	150	"
Blish Milling Co., Seymour, Ind., second order	1 "	250	"
The Cerealine Mfg. Co., Indianapolis, Ind., fourth order	1 "	120	"
The Russell & Miller M. Co., West Superior, Wis	2 "	500	"
R. T. Davis Mill Co., St. Joseph, Mo., second order	1 "	250	"
Metcalf, Miller & Co., Palmyra, Mo	1 "	120	"
J. S. Clark, Troy, Kan	1 "	75	"
Ballard & Ballard, Louisville, Ky	1 "	250	"
Blish Milling Co., Seymour, Ind., third order	1 "	225	"
Taylor Bros., Quincy, Ill., second order	1 "	250	"

2400 H. P. Plant of Heine Boilers at Anheuser-Busch Brewery,
ST. LOUIS, MO.

C. C. Washburn Flouring Mill Co.	3 Boiler,	1000 H.P.
W. R. Klinger, Hermann, Mo.	1 "	100 "
C. C. Washburn Flouring Mill Co., Minneapolis, Minn., second order	3 "	1536 "
V. Bachmann, Indianapolis, Ind	1 "	120 "

BREWERIES AND DISTILLERIES.

Anheuser-Busch Brewing Association, St. Louis, first order	1 Boiler,	200 H.P.
L. Hoster Brewing Co., Columbus, Ohio, first order	1 "	250 "
Hyde Park Brewing Co., St. Louis	1 "	200 "
L. Hoster Brewing Co., Columbus, Ohio, second order	1 "	320 "
Denver Brewing Co., Denver, Colo	1 "	200 "
National Brewery, San Francisco, Cal.	2 "	160 "
J. G. Sohn & Co., Cincinnati, Ohio, first order	1 "	150 "
L. Hoster Brewing Co., Columbus, Ohio, third order	1 "	300 "
Anheuser-Busch Brewing Association, St. Louis, second order	4 "	1200 "
J. B. Wathen & Bro. Co. (distillery), Louisville, Ky., first order,	1 "	300 "
Anheuser-Busch Brewing Association, St. Louis, third order	1 "	200 "
Anheuser-Busch Brewing Association, St. Louis, fourth order	1 "	200 "
Anheuser-Busch Brewing Association, St. Louis, fifth order	4 "	1200 "
J. G. Sohn & Co., Cincinnati, Ohio, second order	1 "	150 "
Fleischman & Co. (distillery), Cincinnati, Ohio	1 "	250 "
San Antonio Brewing Association, San Antonio, Tex	1 "	150 "
Christ Moerlein Brewing Co., Cincinnati, Ohio, first order	1 "	300 "
Cincinnati Brewing Co., Hamilton, Ohio	1 "	200 "
Albert Braun Brewing Association, Seattle, Wash	2 "	240 "
J. B. Wathen & Bro. (distillery), Louisville, Ky., second order,	1 "	300 "
Allen-Bradley Co. (distillery), Louisville, Ky., first order	1 "	250 "
Allen-Bradley Co. (distillery), Louisville, Ky., second order	1 "	250 "
Christ Moerlein Brewing Co., Cincinnati, Ohio, second order	4 "	1200 "
The Central Distilling Co., St. Louis	3 "	900 "
St. Louis Brewing Ass'n, St. Louis	1 "	300 "
Allen-Bradley Co., Louisville, Ky., third order	1 "	300 "
Cincinnati Brewing Co., Hamilton, Ohio, second order	1 "	300 "
Christ Moerlein Brewing Co., Cincinnati, Ohio, third order	1 "	300 "
Standard Brewery, Chicago, Ill	2 "	400 "
Barthomolay Brewing Co., Rochester, N. Y	3 "	750 "
L. Hoster Brewing Co., Columbus, Ohio, fourth order	1 "	300 "
Central Distilling Co., St. Louis, second order	1 "	300 "
Lazcano Y. Gonzalez (distillery), Cardenas, Cuba	1 "	200 "
Wainwright Brewery Co., Pittsburg, Pa.	1 "	250 "
Mutual Distilling Co., Uniontown, Ky	1 "	375 "
Salvador Vidal (distillery), Cardenas, Cuba	2 "	240 "
Mihalovitch, Fletcher & Co., Cincinnati, Ohio	1 "	100 "
The Allen-Bradley Co., Louisville, Ky., fourth order	1 "	80 "
Keystone Brewing Co., Pittsburg, Pa	1 "	250 "
The L. Hoster Brewing Co., Columbus, Ohio, fifth order	1 "	300 "
Beadleston & Woerz Brewing Co., New York, N. Y	2 "	500 "
The L. Hoster Brewing Co., Columbus, Ohio, sixth order	1 "	250 "
M. Winter & Bro., Pittsburg, Pa.	1 "	375 "
R. C. Sibley, East Cambridge, Mass	2 "	500 "
Bay State Distillery Co., East Cambridge, Mass	1 "	500 "
San Antonio Brewing Association, San Antonio, Tex., 2d order,	2 "	300 "
J. Walker Brewing Co., Cincinnati, Ohio	1 "	200 "
L. Hoster Brewing Co., Columbus, Ohio, seventh order	2 "	600 "
American Brewing Association, Houston, Tex	1 "	300 "
Galveston Brewing Co., Galveston, Tex	2 "	500 "
Christian Moerlein Brewing Co., Cincinnati, Ohio, fourth order,	1 "	300 "
Bergner & Engel Brewing Co., Philadelphia, Pa	2 "	500 "
Richmond Brewery, Richmond, Va	1 "	120 "
Anheuser-Busch Brewing Association, St. Louis, sixth order	1 "	300 "

SUGAR PLANTATIONS.

D. H. Cunningham, Sugarland, Tex	1 Boiler,	300 H.P.
J. DeMier, Santa Rosa Plantation, Cuba	4 "	1000 "
J. DeMier, Santa Rosa Plantation, Cuba, second order	2 "	600 "
J. DeMier, Santa Joaquin Plantation, Cuba, third order	2 "	600 "
Casanova Brothers, Carolina Plantation, Cuba	2 "	600 "
Henry Heidegger, Matanzas, Cuba	1 "	250 "
Carlos, Booth & Co.	1 "	60 "
Henry Heidegger, Matanzas, Cuba, second order	1 "	300 "
Casanova Brothers, Carolina Plantation, Cuba, second order	2 "	600 "
Vicente Cagigal & Compartes, Central Gerardo P. Havana, Cuba,	2 "	750 "
J. DeMier, Havana, Cuba, fourth order	1 "	300 "

CABLE AND ELECTRIC STREET RAILROAD COMPANIES.

St. Louis Cable and Western R. R., St. Louis	1 Boiler,	225 H.P.
Central Passenger R. R. Co. (electric), Louisville, Ky., 1st order	2 "	400 "
People's R'y Co. (cable), St. Louis	3 "	600 "
Union Depot R. R. Co. (electric), St. Louis, first order	3 "	750 "
Colorado Springs Rapid Transit R'y Co., Colorado Springs, Colo.,	2 "	300 "
Arlington Heights Electric R'y Co., Fort Worth, Tex	2 "	240 "
Rochester R'y Co. (electric), Rochester, N. Y.	4 "	800 "
Glenwood and Greenlawn Street R'y Co., Columbus, O., 1st order	1 "	120 "
Street Railway Construction Co., Denver, Colo	3 "	375 "
Union Depot R. R. Co. (electric), St. Louis, second order	2 "	500 "
Central Passenger R. R. Co., Louisville, Ky., second order	2 "	400 "
Glenwood and Greenlawn Street R'y Co., Columbus, O., 2d order,	1 "	120 "
City Electric Street R'y Co., Little Rock, Ark	3 "	750 "
Dubuque Electric R'y, Light and Power Co., Dubuque, Iowa	1 "	300 "
Central Passenger R. R. Co., Louisville, Ky., third order	1 "	200 "
Oakland, San Leandro and Haywards Electric R'y Co., Oakland, Cal	2 "	160 "
Broadway and Seventh Ave. R. R. Co. (cable), New York City,	18 "	4500 "
Johnstown Passenger R'y Co., Johnstown, Pa.	2 "	400 "
Union Depot R. R. Co. (electric), St. Louis, third order	2 "	500 "
Salt Lake Rapid Transit Co., Salt Lake City, Utah	1 "	200 "
Oakland, San Leandro and Haywards Electric R'y Co., Oakland, Cal., second order	1 "	140 "
Jersey City and Bergen R'y, Jersey City, N. J	2 "	600 "
Barre Sliding R'y Co., Chicago, Ill	4 "	1500 "
Chicago and North Shore Electric R'y Co., Chicago, Ill	3 "	750 "
Chicago and South Side Rapid Transit Co., Chicago, Ill	2 "	300 "
Altoona and Logan Valley Electric R'y Co., Altoona, Pa	2 "	400 "
Pennsylvania R. R. Co., for Atlantic City Electric R'y, 5th order,	1 "	200 "
Northern Central R'y Co., Canton, Baltimore, Md	1 "	200 "
Ft. Worth and Arlington Heights St. R'y Co., Ft. Worth, Tex., second order	1 "	120 "
Union Depot R. R. Co., St. Louis, Mo., fourth order	4 "	1000 "
East St. Louis Electric Street R'y Co., East St. Louis, Ill	3 "	750 "
Jersey City and Bergen R'y Co., Jersey City, N. J., 2d order	3 "	900 "
Jersey Consolidated Traction Co., Newark, N. J.	1 "	300 "
Jersey Consolidated Traction Co., Newark, N. J., second order,	1 "	300 "
Jersey Consolidated Traction Co., Newark, N. J., third order	1 "	250 "
Allegheny Traction Co., Pittsburg, Pa	2 "	500 "
Hartford Street Railway Co., Hartford, Conn	6 "	2250 "
Union Depot R. R. Co., St. Louis, fifth order	2 "	1000 "
Otis Engineering and Construction Co., Inclined Road, Lake George, N. Y	1 "	160 "
St. Charles Street R'y Co., New Orleans, La	3 "	615 "
Orleans R'y Co., New Orleans, La	2 "	510 "
Louisville R'y Co., Louisville, Ky., fourth order	1 "	200 "
Bergen County Traction Co., Bergen Co., N. J	2 "	500 "
Luzerne, Dallas and Harvey's Lake R'y Co., Wilkesbarre, Pa	3 "	1005 "
Hartford Street R'y Co., Hartford, Conn., second order	2 "	600 "
Lynchburg & Rivermont Street R'y Co., Lynchburg, Va	2 "	400 "
Toledo Traction Co., Toledo, O	2 "	1000 "
J. G. Brill, for Cape Town, Africa	3 "	900 "
Second Avenue Traction Co., Pittsburg, Pa., second order	2 "	740 "
Englewood & Chicago Electric Street R'y Co., Chicago, Ill	3 "	600 "
Fort Pitt Traction Co., Allegheny, Pa	1 "	250 "
Toledo Traction Co., Toledo, O., second order	2 "	1000 "
Philadelphia and Reading Co., Philadelphia, Pa	1 "	425 "
Tamalpais Electric Ry. Co., California	1 "	105 "

RICE AND OIL MILLS.

Howard Oil Co., Houston, Tex., first order	2 Boiler,	500 H.P.
Howard Oil Co., Houston, Tex., second order	1 "	250 "
Galveston Oil Co., Galveston, Tex	2 "	500 "
H. Shumaker Oil Mills, Navasota, Tex	1 "	250 "
Meridian Oil Mills and Mfg. Co., Meridian, Miss	1 "	250 "
Wilmington Oil Mills, Wilmington, N. C	1 "	250 "
Independent Cotton Oil Co., New Orleans, La	1 "	250 "
Capitol City Oil Works, Jackson, Miss., first order	1 "	150 "
Union Oil Co., New Orleans, La	3 "	750 "
Capitol City Oil Mills, Jackson, Miss., second order	1 "	150 "
Crescent City Rice Mill Co., New Orleans, La	1 "	150 "
A. Socola Rice Mills, New Orleans, La	1 "	150 "
Mississippi Cotton Oil Co., Meridian, Miss., second order	1 "	375 "
National Cotton Oil Co., Houston, Tex., third order	1 "	250 "

Station of Edison Illuminating Co.,
ST. LOUIS, MO.
Contains 4500 H. P. Heine Boilers.

National Cotton Oil Co., Galveston, Tex., second order	1 Boiler,	250	H.P.
National Cotton Oil Co., Denison, Tex	2 "	750	"
Union Oil Co., Vidalia, La	1 "	375	"
National Cotton Oil Co., Texarkana, Ark	1 "	375	"
Atlantic Refining Co., Pt. Breeze, Pa	2 "	750	"

ELECTRIC LIGHT, POWER AND GAS COMPANIES.

Springfield Electric Light Co., Springfield, Mo	1 Boiler,	70	H.P.
St. Louis Gas Light Co., St. Louis	1 "	90	"
Evanston Electric Light Co., Evanston, Ill	2 "	140	"
Forest City Electric Light Co., Rockford, Ill	1 "	110	"
Des Moines Edison Light Co., Des Moines, Ia., first order	1 "	200	"
Allegheny County Light Co., Pittsburg, Pa	3 "	960	"
Des Moines Edison Light Co., Des Moines, Ia., second order	1 "	150	"
Columbus Electric Light & Power Co., Columbus, O., 1st order,	1 "	250	"

Station C, Edison Light and Power Co.,
SAN FRANCISCO, CAL.
Contains 1500 H. P. Heine Boilers.

Troy Gas Light Co., Troy, N. Y.	1 Boiler,	80	H.P.
Colorado Electric Co., Denver, Colo., first order	4 "	600	"
Little Rock Electric Light Co., Little Rock, Ark	1 "	200	"
Colorado Electric Co., Denver, Colo., second order	1 "	150	"
Boston Edison Co., Boston, Mass	2 "	500	"
Chicago Edison Co., Chicago, Ill., first order	4 "	1340	"
Brush Electric Light and Power Co., Galveston, Tex	1 "	200	"
Columbus Electric Light & Power Co., Columbus, O., 2d order,	1 "	300	"
Colorado Electric Co., Denver, Colo., third order	1 "	300	"

Company	No.		H.P.	
Columbus Electric Light & Power Co., Columbus, O., 3d order,	1	Boiler,	300	H.P.
Colorado Electric Co., Denver, Colo., fourth order	1	"	150	"
Pueblo Gas and Electric Light Co., Pueblo, Colo	1	"	200	"
Cedar Rapids Electric Light and Power Co., Cedar Rapids, Ia.,	1	"	125	"
Little Rock Electric Light Co., Little Rock, Ark., 2d order	1	"	300	"
Litchfield Electric Light and Power Co., Litchfield, Ill	1	"	150	"
Boston Edison Co., Boston, Mass., second order	1	"	300	"
Laclede Gas Co., St. Louis	3	"	942	"
Hill City Electric Light and Power Co., Vicksburg, Miss	1	"	200	"
University Park Railway and Electric Co., Denver, Colo	1	"	110	"
Hibbard Electric Supply Manufacturing Co., Montreal, Can	1	"	200	"
Columbus Electric Light & Power Co., Columbus, O., 4th order	1	"	300	"
Des Moines Edison Light Co., Des Moines, Ia., third order	1	"	200	"
H. A. & T. C. Gooch, Louisville, Ky	1	"	150	"
Detroit Electric Light and Power Co., Detroit, Mich	2	"	720	"
Chicago Edison Co., Chicago, Ill., second order	4	"	1464	"
City Electric Light Co., Kalamazoo, Mich	1	"	125	"
D. C. Hartwell, Ouray, Colo	1	"	200	"
Milwaukee Power and Lighting Co., Milwaukee, Wis	1	"	320	"
Citizens' Gas Light and Heating Co., Bloomington, Ill	1	"	150	"
Edison General Electric Co., New York	1	"	250	"
Western Electrical Construction Co., Denver, Colo., 1st order	1	"	200	"
Western Electrical Construction Co., Denver, Colo., 2d order	2	"	400	"
Electrical Improvement Co., San Francisco, Cal	2	"	200	"
Gooch Electric Light Co., Louisville, Ky	1	"	256	"
Columbus Electric Light & Power Co., Columbus, O., 5th order,	1	"	300	"
Craig & Son, St. Cunegard, Montreal, Electric Station	1	"	150	"
St. Jean Baptiste Electric Co., Canada, first order	1	"	150	"
Cedar Rapids El. L't and P. Co., Cedar Rapids, Ia., 2d order	2	"	300	"
Ann Arbor T. H. Electric Co., Ann Arbor, Mich	1	"	150	"
Chicago Edison Co., Chicago, Ill., third order	3	"	1125	"
Denver Consolidated Electric Co., Denver, Colo., 7th order	2	"	400	"
Vallejo Electric Light and Power Co., Vallejo, Cal	2	"	150	"
San Francisco Gas Co., San Francisco, Cal	2	"	350	"
Denver Consolidated Electric Co., Denver, Colo., eighth order,	2	"	400	"
Forest City Electric L't and Power Co., Rockford, Ill., 2d order,	1	"	300	"
J. DeMier, Santa Rosa, Cuba	1	"	60	"
Denver Consolidated Electric Co., Denver, Colo., ninth order	1	"	200	"
Brookfield Electric Light Co., Brookfield, Mo	1	"	150	"
St. Jean Baptiste Electric Co., Canada, second order	2	"	500	"
Petaluma Electric Light and Power Co., Petaluma, Cal	2	"	150	"
Edison Electric Illuminating Co., New York	1	"	375	"
Washington Gas Light Co., Washington, D. C	2	"	500	"
Temple Electric Co., Montreal, Canada	2	"	400	"
City Electric Light Co., Kalamazoo, Mich., second order	1	"	150	"
Salt Lake City Gas Co., Salt Lake City, Utah	2	"	750	"
Forest City Electric L't and Power Co., Rockford, Ill., 3d order,	1	"	300	"
Denver Consolidated Electric Co., Denver, Colo., tenth order	1	"	200	"
Siemens & Halske Electric Co., Chicago, Ill	2	"	300	"
Cedar Rapids El. L't and P. Co., Cedar Rapids, Ia., 3d order	1	"	200	"
Edison Electric Illuminating Co., New York, second order	1	"	375	"
Pennsylvania R. R. Co., for Pittsburg U. D. Elec. Light Plant,	2	"	500	"
Chicago Edison Co., Chicago, Ill., fourth order	1	"	500	"
Pennsylvania R. R. Co., for Jersey City Depot Elec. Light Plant,	3	"	1125	"
Mutual Light and Power Co., Montgomery, Ala	3	"	600	"
Toronto Electric Light Co., Toronto, Ont	2	"	520	"
East River Gas Co., Long Island City, N. Y	4	"	600	"
Chicago Edison Co., Chicago, Ill., fifth order	7	"	3500	"
Chicago Edison Co., Chicago, Ill., sixth order	1	"	200	"
Chicago Edison Co., Chicago, Ill., seventh order	1	"	375	"
Chicago Edison Co., Chicago, Ill., eighth order	1	"	500	"
Channon & Wheeler, Quincy, Ill	2	"	620	"
Bridge Mill Power Co., Pawtucket, R. I	2	"	610	"
Chicago Edison Co., Chicago, Ill., ninth order	2	"	1148	"
Woonsocket Electric Machine and Power Co., Woonsocket, R. I.,	2	"	320	"
Dayton Electric Light Co., Dayton, O	3	"	750	"
Denver Consolidated Electric Co., Denver, Colo., 11th order	1	"	375	"
Kalamazoo Electric Co., Kalamazoo, Mich	1	"	375	"
Brookline Gas Co., Boston, Mass	2	"	400	"
The T. Eaton Co., Toronto, Ont	2	"	300	"
Salt Lake & Ogden Gas & Elec. L't Co., Salt Lake City, 2d order,	1	"	375	"
Laclede Gas Light Co., St. Louis, second order	1	"	300	"
A. Von Rosenzweig, Mexico City, Mexico	3	"	375	"
Edison Illuminating Co., St. Louis	15	"	5600	"
General Electric Co., Schenectady, N. Y	1	"	108	"
United Gas Improvement Co., Sioux City, Ia	1	"	255	"

Edison Light and Power Co., San Francisco, Cal., first order ...	4 Boiler,	1500 H.P.
T. Eaton & Co., Toronto, Ont., second order	1 "	150 "
J. J. Vandergrift, Pittsburg, Pa	2 "	500 "
Edison Light and Power Co., San Francisco, second order	2 "	750 "
Brookline Gas Light Co., Boston, Mass	2 "	400 "
Cedar Rapids Elec. Light and P. Co., Cedar Rapids, Ia., 4th order.	1 "	200 "
Indianapolis Gas Co., Cicero, Ind	4 "	1372 "
Logansport & Wabash Valley Gas Co., Windfall, Ind	1 "	343 "
Mutual Light & Power Co., Montgomery, Ala., second order	1 "	200 "
Somerville Electric Light Co., Somerville, Mass	2 "	510 "
New Omaha T. H. Elec. Light Co., Omaha, Neb	1 "	375 "
Chicago Edison Co., tenth order	1 "	574 "
Chelsea Gas Light Co., Chelsea, Mass	1 "	225 "
Salena, Va., Electric Light Plant	1 "	125 "
Newton and Watertown Gas Light Co., Newton, Mass	1 "	305 "
New Omaha T. H. Electric Light Co., Omaha, Neb., 2d order.	1 "	375 "
Citizens Electric Light and Power Co., East St. Louis, Ill	1 "	250 "
Pennsylvania Heat, Light and Power Co., Philadelphia, Pa	2 "	750 "
Toronto Electric Light Co., second order	1 "	250 "

Peoria Water Works,
PEORIA. ILL.
Contains 1200 H. P. Heine Boilers.

WATER WORKS.

Stockton Water Works, Stockton, Cal	1 Boiler,	25 H.P.
Spring Valley Water Co., San Francisco, Cal	3 "	600 "
Houston Water Works, Houston, Tex., first order	2 "	220 "
Lawrence Water Works, Lawrence, Kan	2 "	180 "
National Water Works Co., Kansas City, Mo	4 "	800 "
Texarkana Water Co., Texarkana, Ark	2 "	160 "
Millbury Water Co., Millbury, Mass., first order	1 "	100 "
Millbury Water Co., Millbury, Mass., second order	1 "	85 "
Sheboygan Water Co., Sheboygan, Wis	2 "	150 "
Grafton Water Co., Grafton, Dak	1 "	50 "
City Water Co., Chattanooga, Tenn	1 "	250 "
Norristown Water Co., Norristown, Pa	3 "	300 "

Cincinnati Water Co., Cincinnati, O	2 Boiler,	600 H.P.
Jefferson City Water Co., Jefferson City, Mo	2 "	290 "
Memphis Artesian Water Co., Memphis, Tenn	6 "	900 "
St. Joseph Water Co., St. Joseph, Mo	2 "	260 "
Montreal Water Co., Montreal, Can	3 "	600 "
Houston Water Works, Houston, Tex., second order	1 "	230 "
Peoria Water Works, Peoria, Ill	6 "	1200 "
City Water Co., Chattanooga, Tenn., second order	1 "	250 "
L. & W. B. Bull, Quincy, Ill	1 "	250 "
L. & W. B. Bull, Quincy, Ill., second order	1 "	250 "
Sheboygan Water Co., Sheboygan, Wis., second order	1 "	150 "
Houston Water Works, Houston, Tex., third order	1 "	150 "
St. Joseph Water Co., St. Joseph, Mo., second order	1 "	250 "
H. D. Campbell & Sons, Traverse City, Mich	1 "	150 "
Olympic Salt Water Co., San Francisco	2 "	150 "
Mahanoy City Water Works, Mahanoy City, Pa	1 "	300 "
St. Clair Water Co., Pittsburg, Pa	3 "	480 "
Spring Valley Water Co., San Francisco, Cal., second order	1 "	120 "
Maysville Water Co., Maysville, Ky	2 "	200 "
Worcester Engineering Co., Millbury (Water Co.), Mass	1 "	85 "
Worcester Engineering Co., Millbury (Water Co.), Mass	1 "	85 "
City Water Board, Wheeling, W. Va	2 "	400 "
City Water Board, Wheeling, W. Va., second order	2 "	400 "
Tyler Water Co., Tyler, Tex	1 "	130 "

MINING, COAL AND SMELTING COMPANIES.

Quartz Mountain Mining Co., San Francisco, Cal	1 Boiler,	165 H.P.
Santa Anna Mining Co., Oposura, Mex	2 "	180 "
La Plata Mining and Smelting Co., Leadville, Colo	1 "	110 "
Philadelphia Smelting and Refining Co., Pueblo, Colo	3 "	360 "
Cannon Coal Co., Denver, Colo	1 "	150 "
Alaska Treadwell Gold Mining Co., Douglas Island, Alaska	2 "	400 "
Silver Age Mining Co., Idaho Springs, Colo	1 "	100 "
Boston and Montana C. C. and S. M. Co., Great Falls, Mont	2 "	300 "
Alaska Treadwell G. M. Co., Alaska, second order	1 "	200 "
Madison Coal Co., St. Louis	1 "	200 "
W. Y. O. D. Mining Co., San Francisco, Cal	1 "	150 "
Kilpatrick Bros. & Collins (coal mines), Cambria, Wyoming	1 "	300 "
Magnetic Iron Ore Co., Carthage, N. Y	2 "	300 "
De Lamar Gold Mining Co., De Lamar, Nev	1 "	120 "
Solvay Process Co., Syracuse, N. Y	4 "	1000 "
Hocking Valley Coal Co., Nelsonville, O	2 "	400 "
St. Joe Lead Co., Bonne Terre, Mo., first order	1 "	375 "
Paymaster Mining Co., Ogilvy, Cal	1 "	75 "
St. Joe Lead Co., Bonne Terre, Mo., second order	1 "	375 "
St. Joe Lead Co., Bonne Terre, Mo., third order	1 "	375 "
Chas. Wagner, Mexico	1 "	120 "
Desloge Consolidated Lead Co., Bonne Terre. Mo	2 "	600 "
Omaha and Grant Smelting Works, Denver, Colo	3 "	750 "
St. Joe Lead Co., Bonne Terre, Mo., fourth order	1 "	375 "
Globe Smelting and Refining Co., Denver, Colo	1 "	200 "
Ocean Coal Co., Horatio, Pa	3 "	600 "
Owsley & Cowan, Butte, Mont	2 "	400 "
Arizona Copper Co., Clifton, Ariz	2 "	300 "
De Lamar Gold Mining Co., De Lamar, Nev., second order	1 "	120 "
E. G. Stoiler, Unity Tunnel, Silverton, Colo	1 "	60 "
J. R. De Lamar, Milford, Utah	2 "	250 "
Independence Mine, Victor, Colo	1 "	300 "
Independence Mine, Victor, Colo., second order	1 "	300 "
Anaconda Copper Mining Co., Anaconda, Mont	2 "	750 "
Independence Mine, Victor, Colo., third order	1 "	300 "
Solvay Process Co., Sharon, Pa., second order	2 "	500 "
Anaconda Copper Mining Co., Anaconda, Mont., second order,	2 "	750 "
Mountain Copper Co	1 "	200 "
Canadian Gold Field Co	2 "	160 "
International Coal Mining Co	1 "	150 "

IRON FURNACES.

De Bardeleben Coal and Iron Co., Birmingham, Ala., 1st order,	5 Boiler,	1250 H.P.
Sheffield Iron Co., Sheffield, Ala., first order	3 "	750 "
Lady Ensley Furnace Co.. Sheffield, Ala., first order	3 "	750 "
Mexican Iron Mountain Mfg. Co., Durango, Mex	2 "	300 "
Pulaski Iron Co., Pulaski City, Va., first order	3 "	960 "
Ashland Iron and Steel Co., Ashland, Wis	3 "	450 "
Cameron Coal Co., Cameron, Pa., first order	3 "	750 "

Cameron Coal Co., Cameron, Pa., second order	1 Boiler,	250	H.P
New River Mineral Co., New River Depot, Va	2 "	300	"
Pulaski Iron Co., Pulaski City, Va., second order	1 "	320	"
Sheffield Iron Co., Sheffield, Ala., second order	1 "	250	"
Lady Ensley Furnace Co., Sheffield, Ala., second order	1 "	250	"
De Bardeleben Coal and Iron Co., Birmingham, Ala., 2d order,	7 "	2240	"
Pulaski Iron Co., Pulaski City, Va., third order	1 "	320	"

PACKING HOUSES.

Armour Packing Co., Kansas City, Mo., first order	1 Boiler,	300	H.P.
Armour Packing Co., Kansas City, Mo., second order	1 "	300	"
Kansas City Packing Co., Kansas City, Mo	1 "	300	"
Fort Worth Packing Co., Fort Worth, Tex., first order	1 "	300	"
Fort Worth Packing Co., Fort Worth, Tex., second order	1 "	300	"
Roth-Meyer Packing Co., Cincinnati, Ohio	3 "	300	"
N. K. Fairbank & Co., St. Louis, first order	2 "	600	"
N. K. Fairbank & Co., Chicago, Ill., first order	1 "	500	"
N. K. Fairbank & Co., St. Louis, second order	1 "	500	"
N. K. Fairbank & Co., Chicago, Ill., second order	2 "	1000	"
Nelson Morris & Co., Chicago, Ill	2 "	500	"
K. C. Packing Co., Schwartzschild-Sultzberger Co., 2d order	2 "	600	"
Swift & Co., Kansas City, Mo	2 "	740	"
N. K. Fairbank & Co., Chicago, Ill., fourth order	1 "	250	"
St. Louis Dressed Beef and Provision Co., St. Louis	1 "	375	"
Swift & Co., East St. Louis, Ill., second order	1 "	370	"

MISCELLANEOUS.

Chicago Corset Co., Aurora, Ill	2 Boiler,	120	H.P.
Phœnix Chair Co., Sheboygan, Wis., first order	2 "	450	"
H. E. Roth, Sheboygan, Wis	1 "	30	"
Julius Berkey Felt Boot Co., Grand Rapids, Mich	1 "	140	"
P. B. Mathiason & Co., Bone Black Works, St. Louis	2 "	180	"
W. T. Coleman & Co., Borax Works, San Francisco, Cal	1 "	65	"
California Powder Mills	1 "	65	"
W. S. Townsend, Candy Manufacturer, San Francisco, Cal	1 "	130	"
California Jute Mills, Oakland, Cal	2 "	220	"
M. P. Robinson, Honolulu, Sandwich Islands	1 "	75	"
Springfield Wagon Co., Springfield, Mo	1 "	110	"
G. B. Kane & Co., Chicago, Ill	1 "	20	"
Union Tobacco Works, Louisville, Ky	1 "	70	"
Houston & Texas Central R'y, Houston, Tex., first order	1 "	200	"
J. J. Langles & Co., New Orleans, La	1 "	55	"
Chicago Copper Refining Co., Chicago, Ill., first order	1 "	120	"
R. L. McDonald & Co., St. Joseph, Mo	1 "	80	"
James Roy & Co., Troy, N. Y	2 "	180	"
Chicago Corset Co., Aurora, Ill., second order	1 "	110	"
Hueter Bros. & Co., San Francisco, Cal	1 "	80	"
Houston & Texas Central R'y, Houston, Tex., second order	1 "	200	"
Brittain, Richardson & Co., St. Joseph, Mo	1 "	90	"
Orr's Paper Mills, Troy, N. Y	1 "	125	"
Benecia Agricultural Works, Benecia, Cal	1 "	100	"
A. H. Belo & Co., Galveston, Tex	1 "	50	"
King Kalakua, Honolulu, Sandwich Islands	1 "	80	"
J. G. Johnson & Co., Spuyten Duyval, N. Y	1 "	90	"
Kiddel & Stewart, Denver, Colo	1 "	110	"
Tim Wallerstein & Co., Troy, N. Y.	1 "	50	"
B. J. Johnson & Co., Milwaukee, Wis	1 "	110	"
Carteret Chemical Co., Newark, N. J	1 "	150	"
John Mouat Lumber Co., Denver, Colo.	1 "	110	"
Forest Paper Co., Yarmouthville, Me	2 "	300	"
Cumberland Mills, Cumberland, Me	2 "	300	"
Kansas City, Fort Scott & Gulf R. R., Springfield, Mo	1 "	150	"
Los Angeles Machinery Depot, Los Angeles, Cal	1 "	75	"
A. Bering & Bro., Houston, Tex	1 "	110	"
Wm. H. Bungee, Chicago, Ill	1 "	75	"
Phœnix Chair Co., Sheboygan, Wis., second order	1 "	150	"
Dallas Cotton and Woolen Mills, Dallas, Tex	2 "	400	"
Loomis Gas Machinery Co., Philadelphia	1 "	110	"
Sommer, Richardson & Co., St. Joseph, Mo	1 "	110	"
Charles Stern, Los Angeles, Cal	1 "	165	"
La Constancia Woolen Mills, Durango, Mex	1 "	75	"
A. C. Melchert, Albany, N. Y	1 "	150	"
Arkadelphia Cotton Mills, Arkadelphia, Ark	1 "	150	"
Spring Grove Cemetery, Cincinnati, O	1 "	80	"

Gilbert & Walker, Honolulu, Sandwich Islands	1 Boiler,	55	H.P.
Meier & Kruse, Honolulu, Sandwich Islands	1 "	55	"
Crocker Chair Co., Sheboygan, Wis	1 "	300	"
Mexican Central Railway	2 "	150	"
Christian Peper Tobacco Factory, St. Louis	1 "	200	"
Chicago Copper Refining Co., Chicago, Ill., second order	1 "	100	"
Seeger & Guernsey, N. Y., for Señor Pechado, Toluca, Mex	1 "	75	"
American Mfg. Co., Sheboygan, Wis	1 "	200	"
Gutta Percha and Rubber Mfg. Co., California	1 "	100	"
Williamette Pulp Paper Co., Oregon City, Ore	2 "	300	"
A. Meinicke & Son, Milwaukee, Wis	1 "	150	"
Louisiana Furniture Mfg. Co., New Orleans, La	1 "	150	"
Rockford Hosiery and Mitten Co., Rockford, Ill	1 "	120	"
Beckett Paper Co., Hamilton, O	1 "	250	"
Western Wheel Works, Chicago, Ill	2 "	600	"
A. H. Andrews & Co., Chicago, Ill	1 "	250	"
Publishers, Geo. Knapp & Co., St. Louis Republic, St. Louis	1 "	120	"
Seeger & Guernsey, New York City and City of Mexico, Mex	1 "	80	"
Aug. Beck & Co., Chicago, Ill	1 "	130	"
L. H. Prentice & Co., Chicago, Ill	2 "	150	"
Orrs & Co., Troy, N. Y., second order	2 "	250	"
Tompkins Paper Stock Co., Troy, N. Y	1 "	125	"
Tim Co., Collar and Shirt Factory, Troy, N. Y.	1 "	120	"
Albany Card Paper Co., Albany, N. Y	1 "	150	"
Wm. Angus & Co., East Angus, P. Quebec	1 "	150	"
D. L. Parish Laundry Co., St. Louis	1 "	80	"
Temple Co., Muskegon, Mich	1 "	250	"
Williamette Paper Co., Oregon City, Ore., second order	1 "	200	"
Royal Pulp and Paper Co., East Angus, P. Quebec	2 "	300	"
Western Wheel Works, Chicago, Ill., second order	2 "	400	"
Heath & Milligan Mfg. Co., Chicago, Ill	1 "	200	"
McCormick Harvester Machine Co., Chicago, Ill	1 "	375	"
American Wood Paper Co., Manayunk, Pa	2 "	500	"
Publishers, George Knapp & Co., St. Louis, second order	1 "	120	"
Courier-Journal Co., Louisville, Ky	2 "	240	"
Bausch & Lomb Opt. Co., Rochester, N. Y	2 "	500	"
Denver Paper Mills Co., Denver, Colo	1 "	300	"
Cortina, Pichardo & Co., Toluca, Mex	1 "	75	"
Phœnix Furniture Co., Grand Rapids, Mich	1 "	150	"
Dominion Cotton Mills Co., Canada	3 "	450	"
Robert White & Co., Canada	1 "	150	"
Montreal Carriage Leather Co	1 "	75	"
Sewerage Works, Stockton, Cal	2 "	80	"
Bastion & Valiquetto, Canada	1 "	50	"
Eagle Automatic Can Co., San Francisco, Cal	1 "	100	"
Orrs & Co., Troy, N. Y., third order	3 "	450	"
Bottsford Paper Mill Co., Kalamazoo, Mich	2 "	400	"
Edward E. Barton, Hutchison, Kan	2 "	500	"
Pennsylvania R. R. Co., for Renova, Pa., shops, first order	2 "	600	"
Pennsylvania R. R. Co., for Pittsburg shops, second order	2 "	500	"
Thos. D. Whitaker, Phillipsburg, N. J	1 "	250	"
The Iowa Farming Tool Co., Fort Madison, Iowa	1 "	250	"
W. A. Elmendorf, Chicago, Ill	1 "	20	"
Pennsylvania R. R. Co., for Jersey City, third order	3 "	1125	"
The Burkey & Gay Furniture Co., Grand Rapids, Mich	1 "	300	"
Pennsylvania R. R. Co., Broad St. Station, Phil., Pa., 4th order,	3 "	900	"
Hubbard & Co., Pittsburg, Pa	1 "	250	"
National Starch Mfg. Co., Glen Cove, N. Y	1 "	150	"
Northwestern Terre Cotta Co., Chicago, Ill	1 "	150	"
Buffalo Brass Co., Buffalo, N. Y	1 "	60	"
Mallinckrodt Chemical Works, St. Louis	1 "	375	"
Ferris Wheel, World's Columbian Exposition, Chicago, Ill	3 "	750	"
Smith & Barnes Piano Co., Chicago, Ill	1 "	200	"
Beaver & Co., Soap Works, Dayton, Ohio	1 "	200	"
National Carbon Co., Cleveland, Ohio	4 "	1000	"
L. Waterbury & Co., Brooklyn, N. Y	4 "	800	"
Sterrit & Thomas, Pittsburg, Pa	1 "	50	"
C. L. Colman, Lumber, La Crosse, Wis	1 "	120	"
St. Louis Dried Grains Co	1 "	200	"
Lannett Cotton Mills, West Point, Ga	3 "	900	"
Jno. D. Spreckles Bro., San Francisco, Cal	1 "	30	"
Sterling White Lead Co., Pittsburg, Pa	1 "	250	"
Duryea Starch Co., Glen Cove, Long Island	1 "	200	"
Ætna Paper Co., Dayton, Ohio	1 "	300	"
Wm. Deering & Co., Chicago, Ill	2 "	750	"

Missouri State Penitentiary, Jefferson City, Mo	4	Boiler,	1500 H.P.
Wm. Campbell & Co., New York City	1	"	200 "
Peerless Brick Co., Philadelphia, Pa	1	"	120 "
Whitaker Cement Co., Phillipsburg, N. J., second order	1	"	250 "
National Lead Co., St. Louis	1	"	250 "
Wilmington Cotton Mills, Wilmington, N. C	1	"	250 "
Heath & Milligan Mfg. Co., Chicago, Ill., second order	1	"	200 "
Hamilton Power Co., Montreal, Canada	2	"	240 "
W. G. Warden, Philadelphia, Pa	5	"	750 "
J. L. Ketterlinus, Philadelphia, Pa	2	"	180 "
Theo. Kuntz, Cleveland, Ohio	2	"	400 "
Rockford Mitten and Hosiery Co., Rockford, Ill., 2d order	2	"	400 "
Deering Harvester Co., Chicago, Ill., second order	2	"	928 "

Heath & Milligan Mfg. Co.,
CHICAGO, ILL.
Contains 400 H. P. Heine Boilers.

Leona Cotton Mills, Monterey, Mex	1	Boiler,	100 H.P.
Deering Harvester Co., Chicago, Ill., third order	1	"	250 "
J. Horne & Co., Pittsburg, Pa	2	"	240 "
Pennsylvania R. R. Co., Philadelphia, Pa., sixth order	1	"	500 "
Pennsylvania R. R. Co., Jersey City, N. J., seventh order	1	"	337 "
Partridge & Netcher, Boston Store, Chicago, Ill	1	"	325 "
Mississippi River Dredge Boat "Beta"	4	"	1333 "

Northwestern Terra Cotta Co., Chicago, Ill., second order	1 Boiler,	150	H.P.
R. H. White & Co., Boston, Mass	3 "	630	"
New Orleans Sewerage Co., New Orleans, La	2 "	200	"
National Sewing Machine Co., Belvedere, Ill	1 "	150	"
Ansonia Brass and Copper Co., Ansonia, Conn	4 "	1020	"
Kaufmann Bros., Pittsburg, Pa	1 "	150	"
Griffin Mfg. Co., Griffin, Ga	2 "	300	"
Warren Mfg. Co., Warren, R. I	4 "	1220	"
Eastman's Co., New York City	1 "	250	"
Russell & Co., Massillon, Ohio	2 "	400	"
Burlington Elevator Co., St. Louis	3 "	510	"
Woonsocket Worsted Mills, Woonsocket, R. I	2 "	320	"
Northwestern Terra Cotta Co., Chicago, Ill., third order	1 "	200	"
Arlington Mfg. Co., Arlington, N. J	2 "	250	"
J. W. Peters Fish & Oyster Co., St. Louis	1 "	90	"
Rockford Sugar Works, Rockford, Ill	3 "	900	"
Mallinckrodt Chemical Co., St. Louis, second order	1 "	500	"
New Brittain Knitting Co., New Brittain, Conn	1 "	305	"
Pennsylvania R. R. Co., Philadelphia, Pa., eighth order	2 "	1000	"
Proximity Mfg. Co., Greensboro, N. C	2 "	500	"
Louis Reibold, Dayton, Ohio	2 "	300	"
Fleischmann & Co., Greenspoint, N. Y	4 "	880	"
Frank Jones, Portsmouth, N. H	3 "	600	"
Pennsylvania R. R. Co., Philadelphia, Pa., ninth order	1 "	500	"
Drummond Mfg. Co., Louisville, Ky	1 "	80	"
Sormova Co., Nijni Novgorod, Russia	1 "	250	"
U. S. Dredge Boat Delta.	4 "	1000	"
Wm. A. Talcott, Rockford, Ill	2 "	170	"
Deering Harvester Co., Chicago, Ill., third order	2 "	873	"
Struller, Meyer & Julia Co., City of Mexico	1 "	75	"
R. H. & C. B. Reeves, Camden, N. J	1 "	150	"
S. Ishida, Yokohama, Japan	2 "	500	"
Van Zile & Chrysler, Albany, N. Y	1 "	100	"
Sterling White Lead Co., New Kensington, Pa., second order	1 "	250	"
Chas. F. Joy, St. Louis	1 "	250	"
Booth & Son, California	2 "	200	"
Job Mills, California	1 "	120	"
H. P. Faye & Co., California	1 "	100	"
K. Cohn & Co., California	1 "	105	"
Chelsea Jute Mills, Greensport, N. Y	3 "	1515	"
National Sewing Machine Co., Belvidere, Ill., second order	1 "	250	"

INDEX.

Marquette Building,
CHICAGO, ILL.
Contains 1000 H. P. of Heine Boilers.

HELIOS

www.ingramcontent.com/pod-product-compliance
Lightning Source LLC
Chambersburg PA
CBHW021801190326
41518CB00007B/393

* 9 7 8 3 7 4 3 4 6 7 1 0 1 *